Meteorology

Fourth Edition

Albert Miller

Richard A. Anthes
The Pennsylvania State University

Charles E. Merrill Publishing Company
A Bell & Howell Company
Columbus Toronto London Sydney

Merrill Physical Science Series
Robert J. Foster and Walter A. Gong, Editors

Published by
Charles E. Merrill Publishing Company
A Bell & Howell Company
Columbus, Ohio 43216

This book was set in Times Roman.
The cover was designed by Will Chenoweth.
Cover photo by W. B. Hamilton, U.S. Geological Survey.

Copyright © 1980, 1976, 1971, 1967 by Bell & Howell Company. All rights reserved. No part of this book may be reproduced in any form, electronic or mechanical, including photocopy, recording, or any information storage and retrieval system, without permission in writing from the publisher.

International Standard Book Number, 0–675–08181–5
Library of Congress Catalog Card Number: 79–88079

1 2 3 4 5 6 7 8 9 10 — 85 84 83 82 81 80

Printed in the United States of America

Preface

This is the first revision of *Meteorology* since the death of Professor Albert Miller in 1978. Major changes include the addition of a separate chapter on clouds and precipitation and a section on "forecasting a Christmas snowstorm" in the final chapter. This section integrates many of the concepts presented earlier in the book by showing how they apply to a major winter storm. Material on the origin of the atmosphere, satellites, and air pollution has also been added, and the discussion of extratropical cyclones, fronts, severe thunderstorms, hurricanes, and tornadoes has been updated and expanded. However, much of the material remains intact from the third edition, including atmospheric structure and measurements, atmospheric energy processes and forces, stability of the atmosphere, circulations on all scales, and climate. I have tried to maintain the style of the first three editions while making improvements to the figures and converting to a consistent set of units where necessary.

The contributions of photographs and diagrams by Leo Ainsworth, Ed Brandes, and Joseph Golden are gratefully acknowledged.

Richard A. Anthes

Contents

The Air Around Us

1.1 Introduction

Like a fish in the ocean, man is confined to a very shallow layer of atmosphere; his physical and psychological state—indeed, his very life—depends on his atmospheric environment. Aside from the use of its constituents in biological processes, the atmosphere controls life in many ways. It acts as an umbrella or shield, filtering various types of electromagnetic radiation and high-energy particles from the sun and space. Most meteorites are burned up before they can penetrate to the earth's surface. Winds transport heat and moisture and, in the process, mix the air and create more uniform conditions on the earth than would otherwise exist. The same winds drive the ocean currents, produce waves, erode the soil, and transport pollen and insects. Weather destroys man's structures and disrupts his systems of communication and transportation. The sounds he hears, the scents he smells, and the sights he sees are all affected by the state of the atmosphere.

Meteorology attempts to establish the physical laws or relationships that describe the state of the atmosphere. From these, three practical advantages can be envisioned: (1) prediction of future weather to guide the planning of man's activities, (2) adaptation of man's activities to the atmospheric environment, and (3) modification of weather.

Prediction is a fundamental task of all sciences. Yet, after more than a hundred years of public weather forecasting, forecasts are still the subject of countless jokes. For a period of up to one or two days, the accuracy of weather forecasts is high—though certainly not perfect—but beyond a couple of days, the reliability falls off markedly. Yet meteorologists deal with the same kinds of materials and laws used by other physical scientists. If the astronomer can forecast an eclipse years ahead without a miss, why can't the meteorologist, who has the same basic physical laws at his command, foretell exactly when tomorrow's rain will begin?

We hope the answer to this question will become clear in this book. The complexity of the weather patterns is so great that some meteorologists wonder whether it will ever be possible to completely describe the state of the atmosphere, let alone forecast its future condition in detail. The motions of the atmosphere are composed of convective "cells" and vortices (whirlpools) of many sizes, one superimposed on another. The "chaotic" appearance of a lake or ocean waves on a windy day would be more than equalled in the atmosphere, if air motions could be seen. Yet each whirl plays a role in the total weather picture. It is perhaps not surprising that progress in imposing "order" on the atmosphere so that its behavior can be predicted has been painfully slow.

Existence in harmony with the environment may be natural for most forms of life, but it is not for man with his complex social and technological systems. Climatologists have long been concerned with using knowledge about the atmosphere's characteristics to maximize agricultural production. More recently, with the rapid growth of urban areas and the increased recognition that air is a vital resource, meteorologists have paid greater attention to the interaction of all man's activities and the state of the atmosphere. For example, the level of air pollution is greatly influenced by the state of the atmosphere; but the pollutants in the air may, in turn, affect the weather. Also, our ability to transform large areas of the earth may lead to changes in the weather.

Weather modification has received sporadic, generally minor attention until recently, when attempts to increase rainfall by cloud seeding have renewed interest among both laymen and professional meteorologists. Although the question of the effectiveness of cloud seeding in increasing precipitation is unsettled, we have learned that clouds and fog can be dissipated. There have been numerous other suggestions for controlling the weather or climate, but very few have actually been tried.

This book concerns primarily the weather phenomena that occur in the lowest 10 kilometers (6 miles) of the atmosphere. After this introductory chapter, which deals with the general properties of the atmosphere and measurements, two basic concepts are employed in the discussion of atmospheric processes. One is that the atmosphere is a giant *heat engine*. An engine transforms energy from one type to another. In the atmosphere, radiant energy from the sun is transformed to heat. Because the heat energy of the atmosphere varies from place to place, some of it is changed into kinetic energy; i.e., energy of motion. Man-made engines work the same way, of course. If the gases in the cylinder of a gasoline engine were not hotter than those on the outside, the pistons would not move. Examination of the ways in which different energy levels are created within the atmosphere is a convenient way to decipher the complex processes.

The other concept used in this book is that atmospheric processes and motions exist in a large range of sizes or *scales*. In the case of air motion, for example, there exists a hierarchy of flow systems that range from giant "eddies," which may cover 10% or more of the area of the globe, to tiny whirls, which scatter the dust on a road. Although there is an interplay between each size and its smaller and bigger "brothers," they differ in their characteristics of air motion and weather and in the relative significance of the various atmospheric forces. For example, the circulation pattern of a middle latitude cyclone has a horizontal dimension about a hundred times that of its vertical extent, but in a thunderstorm the depth is about the same as the width. In the case of the cyclone, the earth's rotation is a significant factor in determining the flow. This is not so in the case of the thunderstorm convective cell.

The discussion of atmospheric processes requires so much space that the interesting applications of meteorological knowledge can only be touched on briefly in this book. Except for cloud seeding, which is covered in the second chapter, these topics are left to the final chapter.

1.2 Properties of the Atmosphere

Origin and composition

The atmosphere that exists today evolved slowly over millions of years after the formation of the earth, which occurred approximately 4.5×10^9 years ago. The original gases that formed the early atmosphere were emitted from volcanoes. However, these volcanic gases were considerably different from the gases that constitute our present atmosphere. Table 1.1 shows the composition of gases emitted from present-day Hawaiian volcanoes, while Table 1.2 lists the composition of today's lower atmosphere. Since there is evidence that the composition of the earliest volcanoes was similar to that of present volcanoes, the original atmosphere must have undergone considerable transformation to reach the benign, life-supporting mixture of gases that we have today.

TABLE 1.1 *Percentage by Volume of Gases Emitted by Hawaiian Volcanoes*

Gas	*Percentage*
Water vapor (H_2O)	79.3
Carbon dioxide (CO_2)	11.6
Sulfur dioxide (SO_2)	6.5
Nitrogen (N_2)	1.3
Hydrogen (H_2)	0.6
Other	0.7
Total	100.0

TABLE 1.2 *Composition of Present Atmosphere near Surface*

Gas	*Percentage*
Nitrogen (N_2)	78.08
Oxygen (O_2)	20.95
Argon (A)	0.93
Carbon dioxide (CO_2)	0.03
Water vapor (H_2O)	0.00–3.0 (variable)

Most of the water vapor in the early volcanic eruptions condensed, filling the ocean basins. The carbon dioxide reacted with minerals to form carbonates, while much of the hydrogen escaped the earth's gravitational field. Free oxygen probably formed after the first quarter of the earth's life and after the formation of the first life, which consisted of simple anaerobic plants. These one-celled organisms could produce oxygen through photosynthesis, in which carbon dioxide and water combine to produce carbohydrates and oxygen according to the reaction

$$6CO_2 + 6H_2O \rightarrow C_6H_{12}O_6 + 6O_2 \quad (1.1)$$

It has been estimated that 95 percent of the total oxygen was produced in this way.

Below 80 kilometers, the gases of the atmosphere are relatively well mixed. In this layer, known as the *homosphere,* the proportion of each constituent gas, with few exceptions, is fairly constant throughout. In contrast, in the *heterosphere,* above 80 kilometers, the various gases have tended to stratify in accordance with their weights, as occurs with liquids of different densities.

The lower atmosphere

In addition to the major constituents listed in Table 1.2, a host of other gases such as neon, helium, methane, krypton, xenon, hydrogen and ozone together comprise about a hundredth of 1 percent. The chemical properties of these gases are of considerable interest to the biologist since some, such as nitrogen, oxygen, and carbon dioxide, are involved in life processes. However, the flow of gases into and out of organisms is so slow that it has little effect on the concentration of gases in the atmosphere, and the meteorologist does not normally concern himself with them. Carbon dioxide and ozone are exceptions. As will be explained later, both of these gases play a role in the energy balance of the earth and its atmosphere. Due to differences in rates of production and absorption, the amount of carbon dioxide from place to place varies considerably. For example, over cities where great amounts of fossil fuels such as coal and oil are burned, the concentration tends to be high. There is some speculation that the average CO_2 concentration has been rising due to the increased burning of such fuels during the past half century and that this increase may be changing the atmosphere's heat balance.

Air is never completely dry or pure. There is always some water in the gaseous state, and sometimes it occupies as much as 4 percent of the volume. The amount, however, varies greatly in time and space. Water is the only substance that can exist in all three states — gas, liquid, and solid — at the temperatures that exist normally on the earth. The cycle of transition between these states goes on continuously and plays an important role in maintaining life. In addition, these *phase changes* of water play another role in the atmosphere which is significant to the meteorologist. During the transition from a liquid or solid to a vapor state, water molecules take up some heat energy, which they obtain from the air in which they are contained. When they revert to the liquid or solid state, they release the same amount of energy to their environment. Thus, heat consumed at one place during evaporation may be released at an entirely different place during condensation. This is an effective way of transporting heat over great distances.

Ozone is found in very minute quantities near the surface of the earth, usually comprising less than two parts in a hundred million. If all the ozone in the atmosphere could be brought down to sea level pressure and temperature, it would form a layer only about 2.5 millimeters thick. Although the concentration of ozone is low at all levels of the atmosphere, there is a sharp peak near the altitude of 25 kilometers. Despite the small quantities, ozone is quite significant in the radiant energy transfer that goes on in the atmosphere. Because of its strong absorption of ultraviolet light from the sun, very little of these lethal wavelengths arrive at the surface of the earth. The ozone (O_3) of the atmosphere is believed to form when an atom of oxygen (O), a molecule of oxygen (O_2), and a third "catalytic" particle, such as nitrogen, collide. The atomic oxygen is formed in the atmosphere by the splitting of molecular oxygen under the action of very short wave solar radiation. The maximum of ozone near 25 kilometers is apparently due to a balance of two factors—the availability of very short wave solar energy to produce atomic oxygen, which is gradually depleted as it traverses the upper layers of the atmosphere, and a sufficient density of particles to bring about the collisions required.

Fairly high concentrations of ozone often occur in the lowest few hundred meters of the atmosphere, especially over urban areas. Ozone, which is a corrosive toxic gas, is an important constituent of the so-called "photochemical smog" that afflicts some large cities. The atomic oxygen required for the reaction described above is formed in smog principally through the action of solar radiation on nitrogen dioxide, a product of combustion.

A variety of solid particles are suspended in the air. These include fine dust particles swept up by the wind from exposed soils; soot from forest fires, industrial fires, industrial plants, and volcanoes; pollen and micro-organisms lifted by the wind; meteoritic dust; and salts injected into the atmosphere when ocean spray is evaporated. Large particles are too heavy to remain long in the air, but there are many, so small that they cannot be seen individually with the naked eye, that remain suspended for months or even years. The minute particles of dust thrown high into the atmosphere by the violent eruption of the volcano Krakatoa in the East Indies in 1883 circled the globe for at least two years, producing magnificent sunrises and sunsets.

Many of these small dust particles act as centers, or *nuclei,* around which minute water drops or ice crystals form. (This will be discussed a little later on in Chapter 2, under *Clouds* and *Precipitation.*) Figure 1.1 shows the sizes of these nuclei and, for purposes of comparison, air molecules and liquid and solid water particles.

Dust particles in the air, as well as water droplets and ice crystals, affect the transparency of the air. They not only reduce visibility, but also they prevent some of the sun's energy from penetrating to the surface of the earth. There has been some speculation that humans, who have been increasing the atmosphere's load of dust with factories and high-flying aircraft, may be changing the climate. Although humans are undoubtedly the biggest contributors to the grime in urban areas (where the number of dust particles can reach as high as several million per cubic centimeter in smokeladen air), our contribution to the overall dustiness of the atmosphere is small (at the present time) compared to such natural sources as volcanoes.

The upper atmosphere

The major constituents of the atmosphere remain virtually unchanged up to 80 or 90 kilometers, although there are significant variations in such minor constituents as ozone, dust, and water vapor. But above this level (at which point only

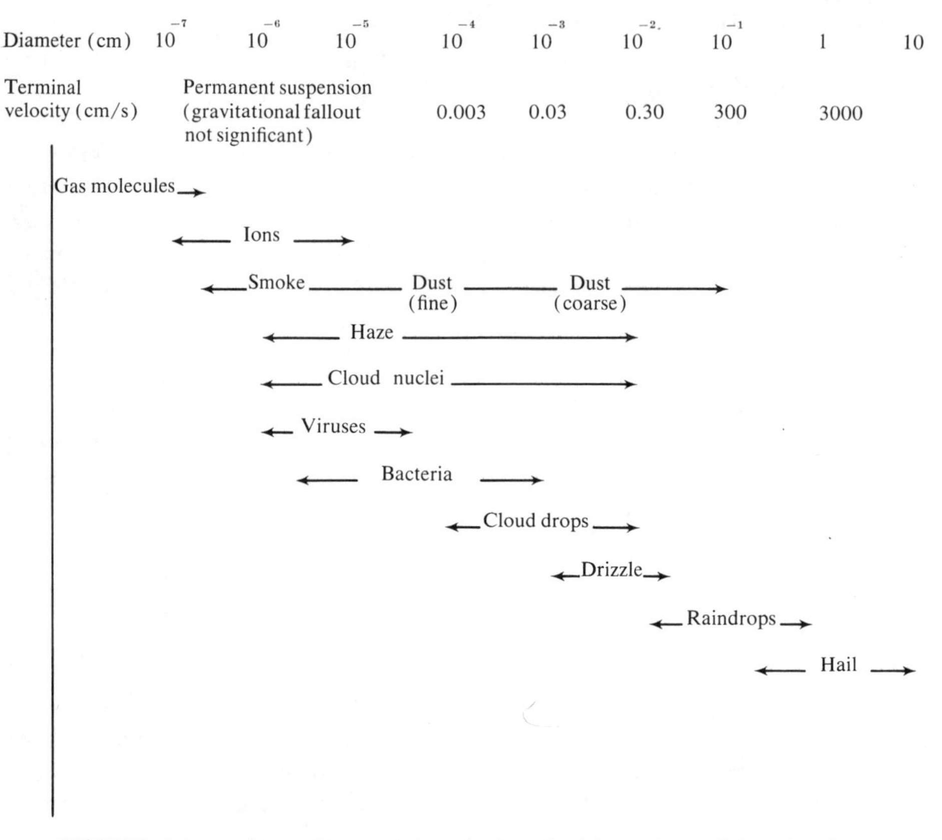

FIGURE 1.1 *Diameter and terminal velocities of particles in the atmosphere.*

0.0002 percent of the total atmosphere remains), the relative amounts and types of gases change. The gases of the "thin" air of the heterosphere undergo various "photochemical" effects induced by the very short waves of ultraviolet and X-ray radiation from the sun. As a result of these photochemical reactions, molecular oxygen is split into two atoms and many molecules and atoms are ionized. That is, electrons have been ejected from their atoms, leaving them with an overall positive charge.

The entire layer from about 80 kilometers upward contains a large number of positively-charged ions and free electrons and is therefore referred to as the *ionosphere.* This electrically charged portion of the atmosphere is very useful for radio communications, since it reflects radio waves. Around-the-world transmissions are accomplished by bouncing radio waves, which move in straight lines, between the ionosphere and the earth's surface.

The distribution of electron density with height fluctuates a great deal. There is a fairly regular daytime-to-nighttime change in the strength of some layers due to the changes in intensity of the solar radiation. In addition, there are occasional sudden ionospheric disturbances (S.I.D.s) and "ionospheric storms" that are associated with disturbances on the sun. An S.I.D. lasts from 15 to 30 minutes, and it is produced by bursts of ultraviolet energy from the sun that cause a sudden increase in the production of electrons. Since electrons absorb part of the radio energy that strikes them, a sudden increase in their number may actually smother the radio energy, leading to fadeouts of communications on the sunlit side of the earth. Ionospheric storms, which can occur during the day or night and last for hours or even days, are believed to be caused by a stream of charged particles emitted from the sun. These fast-moving particles, guided toward the poles by the earth's magnetic field, not only ionize the air, but also they produce the beautiful displays of aurora borealis ("northern lights") and aurora australis ("southern lights").

Temperature distribution in the vertical

The mean temperature distribution in the vertical shown in Figure 1.2 provides a basis for dividing the atmosphere into shells or layers. In the lowest of these layers, the *troposphere,* the temperature decreases with height, on the average, at the rate of 6.5°C/km (3½°F/1000 ft). In this layer, vertical convection currents, induced primarily by the uneven heating of the layer by the earth's surface, keep the air fairly well stirred. Practically all clouds and weather and most of the dust and water vapor of the atmosphere are found in this turbulent layer. Its upper boundary, called the *tropopause,* is at an average elevation of about 10 kilometers, but it varies with time of year and latitude and even from day to day at the same place. Typically, the tropopause is at an elevation of 15 or 16 kilometers over the equator and only 5 or 6 kilometers over the polar regions. It tends to be higher in summer than in winter.

In the *stratosphere,* with an upper boundary at about 50 kilometers, the temperature is first constant. Then it increases with height, reaching a temperature at the *stratopause* that is not much cooler than at sea level. The warmth of this layer is due to the direct absorption of the sun's ultraviolet rays by ozone. It will be shown later that this kind of temperature distribution with height inhibits the air from moving up and down. As a result, this layer acts as a lid on the turbulence and vertical motions of the troposphere. Clouds and vertical convection currents formed near the earth's surface do not usually penetrate very far into the stratosphere. The air in this layer is also quite dry, since it is largely cut off from the sources of moisture at the earth's surface.

The *mesosphere* is the zone between 50 and 85 kilometers in which the temperature decreases rapidly with height, reaching about −95°C at the *mesopause,* which is the coldest point in the atmosphere. Vertical movement of air probably exists in the mesosphere.

Above the mesosphere, the temperature increases rapidly at first, then more slowly, with height. This hot layer is known as the *thermosphere.* At heights above

500–600 kilometers, the density of particles is so low that collisions among them are infrequent, and some of the particles can escape the gravitational pull of the earth. This zone, which marks a transition from the earth's atmosphere to the very thin interplanetary gas beyond, is called the *exosphere*.

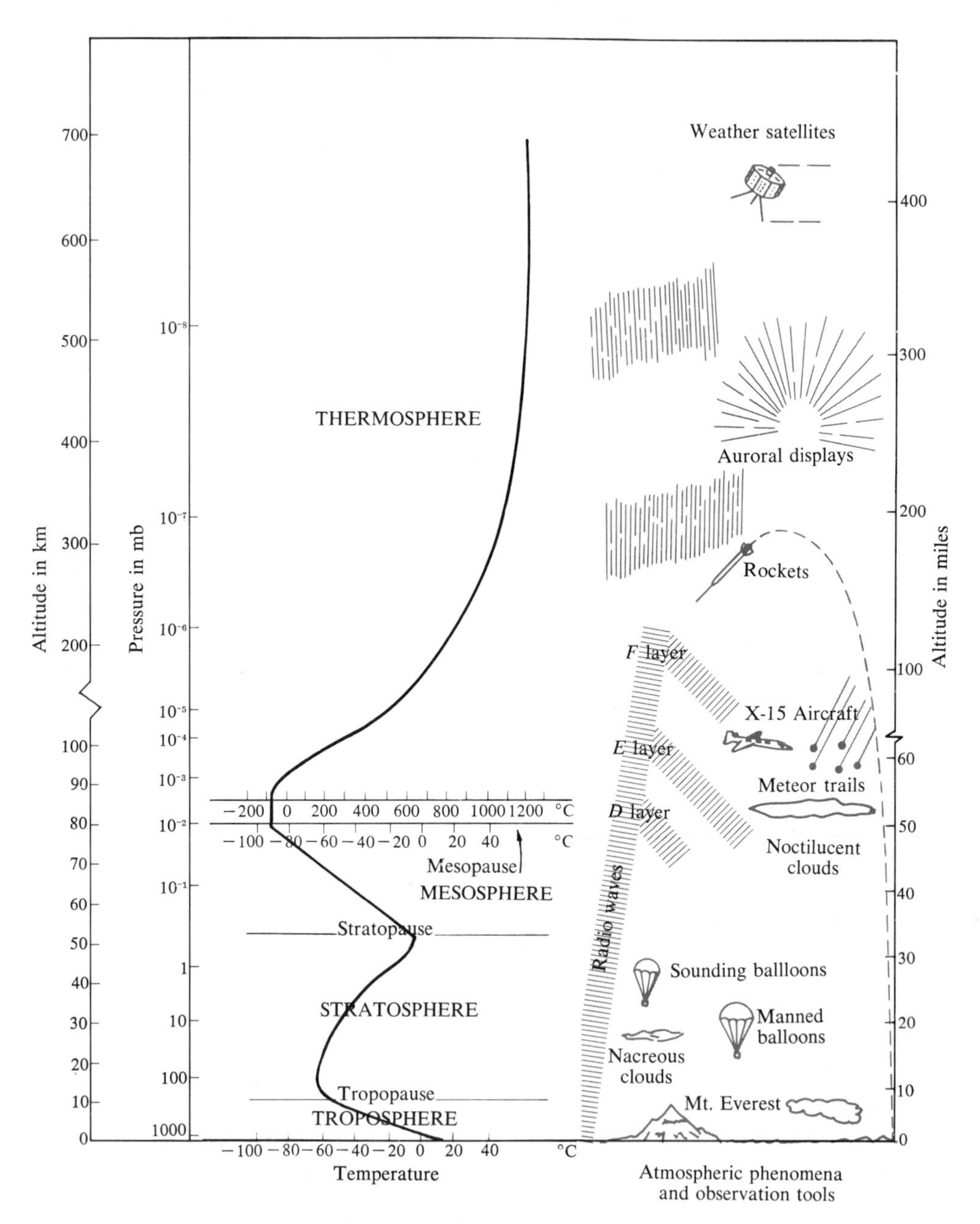

FIGURE 1.2 *Vertical distribution of atmospheric temperature and phenomena.*

1.3 Atmospheric Measurements

Characteristics of the gases

A complete description of the physical state of the atmosphere at any moment requires the measurement of dozens of quantities. This is evident from the discussion of the characteristics of the atmosphere in the last section. There are significant variations in space and time of almost all of the physical properties of the atmosphere. However, some characteristics, such as air temperature, water content, and air motion account for such a large proportion of the total energy and essentially all of everyday "weather" that they deserve special consideration in this chapter. Since the atmosphere is composed mostly of gases, we shall briefly review the behavior of gases and the measurements that are used to describe their physical state.

First, matter can exist in any of three states — solid, liquid, or gas. Almost all substances on the earth occur naturally in only a single state. The gross characteristics of each of the three states are quite familiar. Solids resist changes in their shape and do not flow, while liquids and gases are easily deformed and do flow (and thus are known as *fluids*). The space occupied by a solid or a liquid is not easily altered, but a gas spreads out to fill the entire volume available to it. (We say that gases are *compressible.*)

These characteristics can be explained largely in terms of how closely bound the molecules of the substance are. In a solid, the molecules are locked into position and their motion is restricted to oscillations over short distances from their mean positions. The molecules in a liquid have considerable freedom of movement, but they are bound to the bulk of the liquid with sufficient force so that they cannot significantly increase the mean distance between individual molecules. In a gas, the *adhesion* between molecules is weak, the molecules are relatively far apart, and they can move with comparative freedom throughout the volume occupied by the gas.

We shall be concerned primarily with the "gross" properties, particularly those of gases. Although, we may occasionally mention molecules and atoms in the way of explanation, we shall be interested primarily in populations of molecules and atoms and how they behave as groups.

One such gross property of a gas is the space that a given number of gas particles may occupy. At sea level there are about 25×10^{18} molecules of air in each cubic centimeter (about the volume of the tip of your small finger up to the base of the nail). The total mass of molecules per unit volume is the density. Near sea level, 1 cm^3 of air contains a mass of about 1.2×10^{-3} g, so the density is 1.2×10^{-3} g/cm^3.

Another bulk property of fluids is *pressure,* which is defined as the force per unit area exerted on any surface being bombarded by the fluid's moving molecules. Held to the earth by gravitational attraction, the earth's atmosphere has a cumulative force or weight per unit area averaging 14.7 lb/in^2 at mean sea level. In the SI system of units* the standard sea level pressure is 101.3 kilopascals (kPa), where 1 Pa is 1 $kg/m/s^2$. For many meterological purposes, including weather

*Système International d'Unités (designated SI in all languages).

reports, the millibar (mb), which is 0.1 kPa, is the commonly used unit of pressure. Thus the average sea level pressure is 1013 mb.

Gases are easily compressed, a fact that is illustrated by the pressure distribution with height given in Figure 1.2 and in the table of Appendix 2. In water, the pressure increases almost exactly in proportion with depth below the water surface, but not so in the atmosphere. As can be seen from the table, one must ascend 2500 meters for the pressure to fall off 25 percent (about 250 millibars) from its sea level value, but over 3000 meters more for another 25 percent drop, and another 4800 meters for an additional 25 percent. In the layer between sea level and 5000 meters, the pressure decreases 470 millibars, but in the layer between 30,000 meters and 35,000 meters, the decrease is less than 7 millibars. Evidently, the air near the bottom of the atmosphere is compressed by the weight of the air resting above. The compressibility characteristic of gases is of great significance in atmospheric processes. As will be pointed out in a later chapter, rapid compressions and expansions of air occur naturally in the atmosphere due to vertical displacements, and these are largely responsible for much of the weather.

Gas laws

Two experimental laws relate the properties of temperature, density (or volume), and pressure in gases. (1) Boyle's law states that if the temperature of a gas does not change, its pressure varies directly as the density varies. In symbolic form, $p = k\rho$, where p is the value of the pressure, ρ is the density, and k is a constant. (2) Charles's law states that if the pressure within a gas is kept unchanged, its volume will change in proportion to any temperature change that may occur. This means that, for a fixed amount of mass, the density (mass per volume) is inversely proportional to the temperature when the pressure is constant; i.e., the density decreases when the temperature increases, and it increases when the temperature decreases. At the same pressure, cold air is denser than warm air.

Boyle's law and Charles' law may be combined to relate pressure (p), temperature (T), and density (ρ) through the *equation of state*

$$p = \rho RT \tag{1.2}$$

where R is the gas constant for dry air. The value of R is 287 joules per kilogram per kelvin (J/kg/K).*

Observations of the atmosphere

Weather observations of a sort have been made by man since earliest times, but systematic measurements of the elements did not begin until the invention of instruments during the seventeenth and eighteenth centuries. Until the twentieth century, measurements were confined to the air close to the ground. Systematic measurements of most of the earth's atmosphere are scanty even today.

Measurement of the state of the atmosphere is quite difficult. In addition to the usual requirement that an instrument measure accurately whatever it is designed

*A joule is a unit of energy and is equal to 4.187 calories.

to measure, the meteorological instrument must be rugged enough to withstand the weather elements—the force of buffeting winds, the corrosive action of high humidity and flying dust, the extremes of heat and cold. Another difficulty is the inaccessibility of much of the atmosphere, so that instruments must be built to transmit their measurements to distant ground points; they must be rugged and light enough to be carried aloft by balloons and rockets and cheap enough so they can be used in the large quantities necessary to observe the atmosphere. Finally, meteorological measurements must be "representative," a difficult objective to achieve, considering the enormous size of the atmosphere and the comparatively few observations that can be made. Taking the depth of water in a single 8-inch diameter rain gauge as representative of the average rainfall over an area of many square miles is somewhat like assuming that the height of a single student taken at random is equal to the average of the entire school.

Temperature

A thermometer is a device that measures the degree of hotness or coldness of a body on a numerical scale. This is usually done by correlating physical changes in the thermometer with temperature changes. Thus, for example, the increase in volume of mercury with increased temperature is used in the common mercury-in-glass thermometer. Another way to measure the temperature of a body is to relate changes in the properties of the body itself with changes in its temperature. The color of steel in a furnace is a good index of its temperature; the speed at which sound waves travel through air depends on the air temperature.

Almost every type of temperature measuring device has been used in meteorology, but the expansion type is the most commonly used for observation near the surface of the earth because of its ruggedness and cheapness. The ordinary liquid-in-glass thermometer is widely used in meteorology. Liquid-in-glass thermometers that will register the maximum or minimum temperature during a period requires slight modifications. The maximum thermometer has a constriction of the bore of the glass tube just above the bulb; as the temperature rises, the mercury is forced through the constriction, but when the temperature falls, the weight of the mercury in the column is insufficient to reunite it with that in the bore. Thus, the top of the mercury column indicates the highest point reached. The maximum thermometer can be reset by shaking the thermometer, thereby forcing the mercury through the constriction.

The minimum thermometer contains alcohol in the bore, with a small, dumbbell-shaped glass index placed inside the column of alcohol. The index is kept just below the meniscus of the alcohol column by surface tension. With the thermometer mounted horizontally, when the alcohol contracts, the meniscus drags the index with it; but when the alcohol expands, the meniscus advances, leaving the index at its lowest point. To reset, the index can be returned to the meniscus merely by tilting the thermometer.

Expansion-type thermometers are also used for recording the temperature. Either a bimetal or a Bourdon thermometer is used to move a pen arm that traces its position on a paper chart driven by a clock. The bimetal thermometer is the type ordinarily used in thermostatic control devices. Two strips of metal, having different rates of expansion during a temperature change, are welded and rolled together. The difference in expansion of the two strips causes changes in the curvature of the element as the temperature changes; with one end fixed in position, the other end is free to move the pen arm and indicate the temperature. The Bourdon thermometer consists of a flat, curved metal tube containing a liquid. As the volume of the liquid changes with temperature, the curvature of the tube changes, and this can be used to move the pen arm in a fashion similar to that of the bimetallic strip.

The principal use of electrical thermometers in meteorology is in the radiosonde, which is attached to a balloon and transmits temperature, pressure, and humidity data by radio as it ascends through the atmosphere. There are two general types of electrical thermometers: (1) the thermoelectric thermometer, which operates on the principle that temperature differences among the junctions of two or more different metal wires in a circuit will induce a flow of electricity; and (2) the resistance thermometer, which is based on the principle that the resistance to the flow of electricity in a substance depends on its temperature. It is the latter that is used in the radiosonde. The ceramic elements commonly used are called thermistors.

Obtaining meaningful air temperatures is not a simple procedure. Air is a poor conductor of heat and quite transparent to radiation, especially in the short wavelengths emitted by the sun. That this is so is quite evident when one moves a few feet from the shade into the sun; even through the air temperature is almost identical, one feels much warmer in the sun. Or, standing near a fire, the side of the person facing the fire "roasts" while the other side "freezes."

Most thermometers are much better radiation absorbers than air. They absorb energy from the sun and other warm objects that passes right through the air. If the thermometer is to measure the air temperature, such radiation must be prevented from reaching the thermometer. This is accomplished by shielding the thermometer, at the same time keeping it in contact with the air. This can be done by enclosing the thermometer within a highly polished tube, allowing plenty of room for air to circulate past the thermometer. However, when several temperature measuring devices are used, it is convenient to house them in a special "instrument shelter," which permits air to pass through. The shelter also keeps the instruments dry during rain. This is important because a wet thermometer will generally read lower than a dry one. (See the section on humidity.)

The thermometer should be ventilated artificially when there is little wind because the conductivity of air is poor ("dead," or stagnant, air is often used for insulation), and the thin layer that encases the thermometer might have a different temperature than the "free" air. By stirring the air, this layer is mixed with the surrounding air.

Even if the precautions in measuring temperature given above are taken, there is still the question of how to interpret temperature measurements. On a sunny, windless day the temperature of air within an inch or two of a cement sidewalk can be 30°F or more higher than at the 4-foot level. Even at the same height above the ground, the temperature differs greatly between the city and the country, within forests and over open land, along sloping land and flat land. Differences in the thermometer environment are so important that measurements at a single point can rarely be considered representative of the average conditions closer than 2°F.

Pressure

Pressure is defined as the force per unit area exerted on any surface in a fluid. The orientation of the surface will not affect the pressure. In the case of the atmosphere, which has no outer walls to confine its volume, the pressure exerted at any level is due almost entirely to the weight of the air pressing down from above; i.e., the force results from gravitational attraction. (The units of pressure were given in the section on characteristics of gases.)

As was explained earlier, the pressure changes most rapidly in the vertical. In the lowest few kilometers, the pressure decrease amounts to about 1 millibar per 10 meters. Because of the compressibility of air, the rate at which the pressure decreases with height becomes slower at greater heights.

Variations of pressure in the horizontal are much smaller than they are in the vertical. Near sea level, the change of pressure with distance rarely exceeds 3 millibars per 100 kilometers and is usually much less than half this rate. Although the horizontal variations in pressure are small, they are responsible for the winds we observe. Since pressure measures the weight per unit area of the atmosphere, variations in pressure along any horizontal surface (such as sea level) must arise through variations in the average density of the atmosphere; i.e., there must be more molecules in a column of air above a point having high pressure than in one above a point where low pressure is observed.

Pressure also changes with time at a single place. Some of these changes are of an irregular nature, caused by occasional invasions of air having a different mean density. But there is also a quite regular diurnal oscillation of the pressure that causes, on the average, two peaks (at about 10 A.M. and 10 P.M.) and two minima (at about 4 A.M. and 4 P.M.). The difference between maxima and minima is greatest near the equator (up to 3 millibars), decreasing to practically zero in the polar regions. These regular fluctuations in the pressure are analogous to the tidal motions in the ocean, but in the case of the ocean it is the gravitational pull of the moon and sun that causes the water surface to bulge slightly outward from the earth, while in the atmosphere the daily heating and cooling cycle appears to be the dominant cause of the pressure variations. Diurnal wind oscillations accompany the migration of these maxima and minima of pressure around the earth each day; these are hardly detectable at low elevations in the atmosphere, but they become quite strong between 80 and 100 kilometers.

The mercurial barometer, invented by Torricelli in 1643, is still the fundamental instrument for measuring atmophere pressure. It is constructed by filling a long glass tube sealed at one end with mercury, inverting the tube, and placing the open end into a dish of mercury. The mercury in the tube will flow into the dish until the column of mercury is about 30 inches high (at sea-level site), leaving a vacuum at the top. In principle, the barometer is merely a weighing balance (Figure 1.3), the presence exerted by the atmosphere on the exposed surface of the mercury in the dish equaling that exerted by the mercury in the tube. Changes in atmospheric

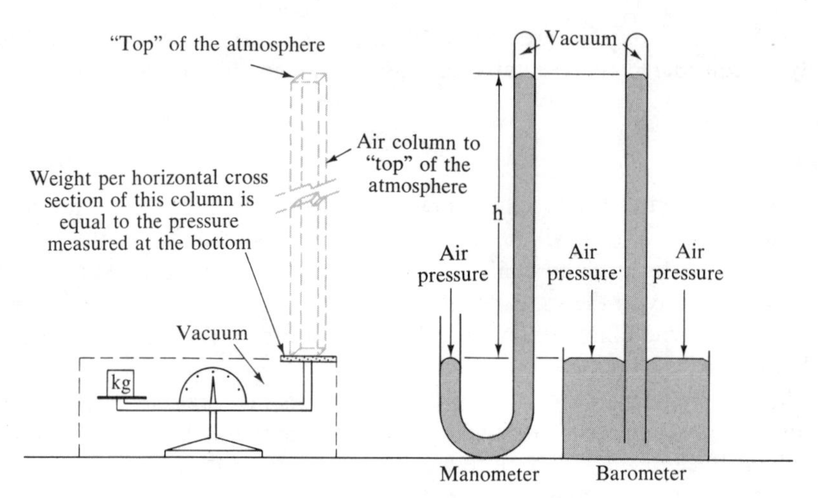

FIGURE 1.3 *Principle of the barometer.*

pressure are detected from changes in the height of the column of mercury. Although it is now the custom to use the height of the column as a pressure unit (millimeters or inches of mercury), conversion to such units as millibars can be made as follows.

The density of mercury (Hg) at 0°C is 13.6×10^3 kg/m^3. (Note that the height of the column of mercury will depend on temperatures as well as pressure, since mercury expands with increased temperature. To obtain the true pressure, one must correct for this mercury expansion. The mass of a column of mercury = mercury density × volume = density × height × cross-sectional area of tube. Its *weight,* therefore, would be obtained by multiplying by the acceleration of gravity (weight = mass × gravity) and the weight per unit area (pressure) exerted by the column obtained by dividing by the area. Thus, pressure = gravity × density × height. For example, if the height of the column of mercury were 76 centimeters (0.76 m), the pressure p would be

$$\begin{aligned} p &= (9.8 \text{ m/s}^2) \times (13.6 \times 10^3 \text{ kg/m}^3) \times (0.76 \text{ m}) \\ &= 101.29 \times 10^3 \text{ kg/m/s}^2 \\ &= 101.29 \text{ kPa} \\ &= 1012.9 \text{ mb} \end{aligned} \tag{1.3}$$

Here a value of 9.8 m/s^2 has been used for gravity. In practice, since gravity varies slightly, the value at the particular place should be used.

The aneroid barometer, although not usually as accurate as the mercurial barometer, is more widely used because it is smaller, more portable, usually cheaper to manufacture, and simpler to adapt to recording mechanisms. Its principle of operation is that of the spring balance (Figure 1.4). A thin metal chamber, with most of its air evacuated, is prevented from collapsing under the force of atmospheric pressure by a spring. The force exerted by the spring depends on the distance it is stretched. The balance between the spring force and the atmospheric force will thus

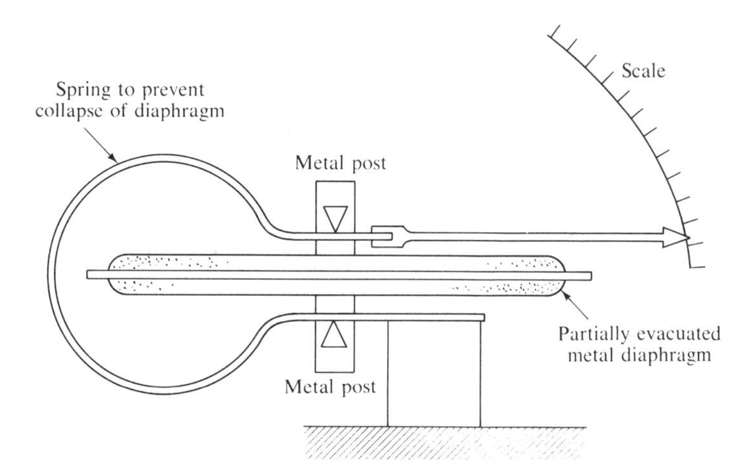

FIGURE 1.4 *Principle of the aneroid barometer.*

depend on the width of the chamber. Changes in this width can be discerned by movement of an arm attached to one end of the chamber; these deflections are usually magnified by levers. If a pen is attached to the arm, the instrument becomes a barograph.

The altimeters used in aircraft and by mountain climbers, surveyors, and others are usually nothing more than aneroid barometers made to indicate altitude rather than pressure. They are designed to give the altitude for the standard ("normal") pressure distribution with height, and so will give slightly erroneous readings. For accurate determinations of altitude, the true density of the air for each altitude increment must be measured and corrections to the indicated altitude must be computed.

Humidity

The concentration of gaseous water in the atmosphere varies from practically zero to as much as 4 percent (4 grams of water in every 100 grams of air). The extreme variability in the amount of water vapor in both space and time is due to water's unique ability to exist in all three states—gas, liquid, and solid—at the temperatures normally found on earth. Water vapor is continuously extracted from the atmosphere through condensation (vapor to liquid) and sublimation (vapor to ice); some of this may fall to the earth through precipitation. Water is continuously being added to the atmosphere through evaporation (liquid to vapor) from oceans, lakes, rivers, soil, plants, and raindrops, and also sublimation (ice to vapor) from snowflakes, glaciers, etc.

The exact amount of water vapor that exists at any place and time is important to the meteorologist because of the role water plays in weather processes. First, it is significant because condensation is an important aspect of "weather." Second, water vapor is the most important radiation absorber in the air and thus affects the energy balance of the atmosphere (Chapter 3). Third, the release of the latent heat of condensation is an important source of energy for the maintenance of atmospheric flow.

For a substance such as water to change its phase from solid to liquid or liquid to gas, the forces that bind the molecules together must be broken down. Work must be done in overcoming these intermolecular forces, so the molecules must expend part of their internal energy. The molecules acquire this energy from their environment. It is for this reason that skin is cooled by evaporation of perspiration and water in a porous water bag is cooled by evaporation through the walls. The energy required to effect a "phase," or state, change such as occurs in evaporation is called *latent heat* because it reappears when the change of state is reversed. Thus, to evaporate 1 gram of liquid water, approximately 600 calories are required. If the same gram of gaseous water is returned to liquid state, the 600 calories will be released to the environment. A similar thing happens during the ice to liquid transition, but the *latent heat of fusion* is only about 80 cal/g.* Changes directly between ice and vapor involve a latent heat of sublimation which is the sum of the latent heats of fusion and vaporization; i.e., approximately 680 cal/g.

We refer to the gaseous state of water as vapor because it is so easily condensed, but it acts much like any other gas in the atmosphere. The molecules of water vapor move about, occupy space, and exert pressure as do the other gases, except that the amounts of the other gases are relatively fixed. The quantity of water vapor in the air can be expressed in a variety of ways. One is the density of water vapor, usually referred to as the *absolute humidity* and expressed as the number of grams of water vapor in a given volume. Normally, there are not more than about 12 g/m^3, although as much as 40 g/m^3 can occur.

The *partial pressure* of water vapor (i.e., the contribution made by water to the total atmospheric pressure) is another measure that can be used. It is usually expressed in millibars or inches of mercury. Typically, the water vapor pressure does not exceed 15 millibars, although it can reach double or more this value.

The amount of water vapor that can be added to a volume at any given atmospheric pressure and temperature is limited. When a volume has reached its capacity for water vapor, it is said to be saturated and the volume will accept no more gaseous water. The *saturation vapor pressure,* as this maximum vapor pressure is called, is a function of temperature. (See Figure 1.5.) This dependence of the saturation vapor pressure on the temperature is why cooling is so important in producing condensation. For example, if a sample of air having a temperature of 70°F and a water vapor pressure of 20 mb were cooled to 45°F, the sample would become saturated at a temperature of 64°F and further cooling would result in condensation of the excess moisture. When the 45°F temperature were reached, the saturation vapor pressure would be only 10 mb; so half of the vapor would have liquefied. The temperatue to which a sample of air must be cooled (at constant atmospheric pressure) to make it "saturated" is called the *dew point.* In the above example, the dew point of the sample before condensation began was 64°F; after condensation begins, the temperature and the dew point are equal. Thus, the dew point is a direct measure of the water vapor pressure; the difference between the temperature and the dew point is a measure of the degree of saturation of the air.

---

*The latent heats depend somewhat on the temperature at which the phase changes occur.

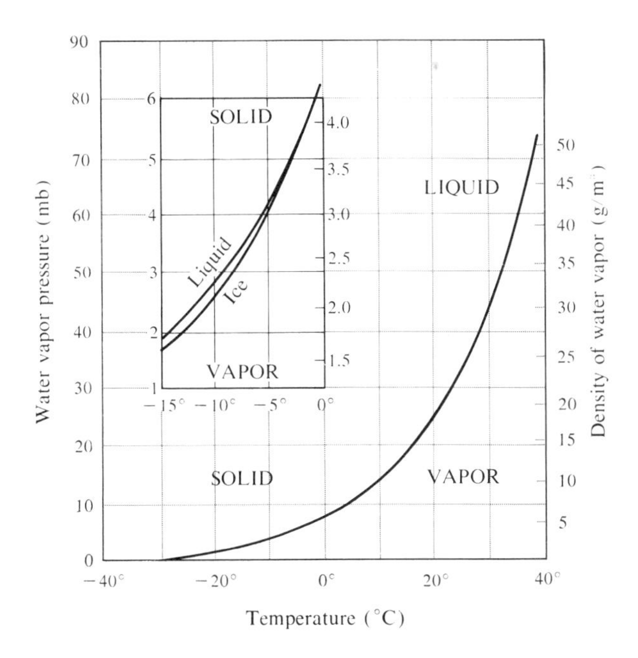

FIGURE 1.5 *Saturation vapor pressure and density as a function of temperature. (Insert shows variation below 0°C.)*

The *relative humidity* is the ratio of the actual vapor pressure to saturation vapor pressure; i.e., relative humidity = actual vapor pressure/saturation vapor pressure. The ratio is usually multiplied by 100 and expressed in percent. Relative humidity measures how close the air is to saturation (100 percent indicates complete saturation). In the example given above, the relative humidity before cooling was $20/25 \times 100 = 80$ percent; between 64°F and 45°F, the relative humidity remained constant at 100 percent. The relative humidity is very sensitive to temperature change. There is normally a large diurnal change of relative humidity, even when the quantity of moisture in the air is constant, merely because the daily temperature variation continuously changes the saturation vapor pressure.

The most accurate way to measure humidity is to extract all of the water vapor from an air sample (perhaps by passing it through a chemical drying agent) and then weigh the water collected. However, for meteorological observations, this procedure is not practical, since the sampling time is too long and the analytical tools are too complicated. Study of the atmosphere requires almost instantaneous sampling under field conditions.

None of the many techniques used to measure atmospheric humidity are completely satisfactory; here we will mention just a few of the most commonly used instruments. The hair hygrometer is probably the oldest and most widely used instrument for measuring moisture. Many organic materials such as wood, skin, and hair absorb moisture when the humidity is high, and so they expand. Blond human head hair increases its length by about 2½ percent as the relative humidity increases from 0 to 100 percent. The hair hygrometer merely consists of one or more hairs whose changes in length are made to move a pointer or, in the case of a hygrograph, a pen.

The psychrometer consists of a pair of ordinary liquid-in-glass thermometers, one of which has a piece of tight-fitting muslin cloth wrapped around its bulb. The cloth-covered bulb, called the wet bulb, is wetted with pure water, and both thermometers are then ventilated. The dry bulb will indicate the air temperature, while the wet bulb will be cooled below the dry bulb temperature by evaporation. The amount of evaporation, and therefore of cooling, will depend on how nearly saturated the air is. If the surroundings are saturated, there will be no evaporation, and the wet and dry bulbs will read the same. The difference between the dry and wet bulbs, called the *depression of the wet bulb,* is a measure of the degree of saturation of the air. Tables can be used to obtain the relative humidity or dew point from psychrometric readings.

An electrical hygrometer is used in the radiosonde. It consists of an electrical conductor that is coated with lithium chloride, which is hygroscopic. The amount of moisture absorbed by the conductor depends on the relative humidity of the air, and the electrical resistance of the conductor is a function of its dampness.

Wind

Air in motion, or *wind*, is the "equalizer" of the earth's atmosphere. By transporting heat, moisture, pollutants, etc., from one place to another, it acts to redistribute the concentration of these quantities. Chapter 4 will discuss what causes winds; at this point, only some of the general characteristics of wind will concern us.

Although air moves up and down as well as horizontally, the speed of vertical displacements is usually a tenth or less of the horizontal component. Even though it is quite small, the vertical component is very important. As we shall see later, it is the up-and-down motion of air that is principally responsible for the formation and dissipation of clouds in the atmosphere. Only the *horizontal* component of the wind is measured on a regular basis, while the much smaller vertical component must be computed from relationships between it and the changes of the horizontal wind in space.

Many instruments are used to measure the horizontal wind velocity near the surface of the earth. The wind vane is a very old device for indicating wind direction. Because it points *into* the wind, it is customary to designate wind direction as that *from* which the air comes. Thus, when the air is moving *from* northwest (315°) to southeast (135°), the wind direction is said to be northwest (315°).

A large variety of *anemometers* exists for measurement of wind speed. The cup anemometer is probably the most widely used. It consists of three or more hemispherical cups clustered around a vertical shaft. Air striking concave sides of the cups exerts more force than that hitting the convex sides, causing the cups and therefore the shaft to turn. The number of rotations per unit time is a measure of the wind speed.

Air flow is retarded by friction with the ground and deflected by obstacles, so that the position of wind-measuring instruments must be carefully considered. At an airport, for example, both the wind speed and direction atop the control tower may be considerably different from those at the end of the runway. Typically, the wind speed increases rapidly with height near the surface, so that the height of an anemometer will greatly influence the speed recorded. Unfortunately, there is no uniformity of height for anemometers, although arbitrary standards have been set.

Gustiness and the diurnal wind variation Anyone who has watched a wind vane oscillate and a cup anemometer alternately increase and decrease its rotation speed during brief periods of time, or who has watched a flag flutter in the wind, can attest to the normal unsteadiness of the wind. An example of such fluctuation can be seen from the recording of the wind direction and speed of Figure 1.6. These velocity changes are attributed to the fact that air normally does not move in straight lines, but in tortuous paths. We call such erratic air flow patterns turbulent. Successive particles passing a single point in space may have had distinctly different histories, some having been most recently at higher elevations than the point, others at lower elevations. Normally, those coming from higher up will arrive with relatively high velocities, while those coming from lower elevations will have relatively low velocities. The result will be a gusty wind velocity at the observation point.

On clear nights strong radiative cooling at the ground often produces a layer of very cold air next to the ground which is overlain by warmer air aloft (an inversion). Under such conditions the air near the surface is very stable, vertical mixing is inhibited, and the surface wind is often nearly calm or slows a light flow from a direction determined by local effects such as sloping terrain. The light southerly winds between 9 and 10 A.M. in Figure 1.6 is an example of the winds under such stable conditions. Shortly before 10 A.M. on this day, surface heating destroyed the inversion and air from higher aloft was mixed down to the surface. This larger-scale current of air was predominantly from the west, and so the surface wind direction shifted when the inversion was eliminated. As the vertical stability is reduced by surface heating, the winds become more turbulent. Therefore, strong, gusty winds are most likely to occur near midday when the solar heating is a maximum.

Upper-air observations

Sounding techniques Winds at levels above the reach of groundbased instruments are measured by tracking helium-filled or hydrogen-filled balloons. The horizontal displacements over short intervals of time as the balloon ascends give the velocity. The changes in position of the balloon may be determined by any of these methods: (1) optically, by the use of a theodolite (similar to a surveyor's transit); (2) by reflection of radio waves (radar) from a target carried by the balloon; (3) by tracking of the radio signal transmitted by a radiosonde carried by the balloon.

Systematic measurements of meteorological conditions in the free atmosphere high above the surface began around the turn of the century. Until 1938, when the radiosonde came into use, the sounding instruments had to be retrieved before the data became available. Instruments recording temperature, humidity, and pressure were carried aloft by balloons, kites, and airplanes. In the case of a free balloon, the instrument dropped to the earth on a parachute, and the processing of the data had to wait until some finder returned the instrument. Kites were extremely laborious to

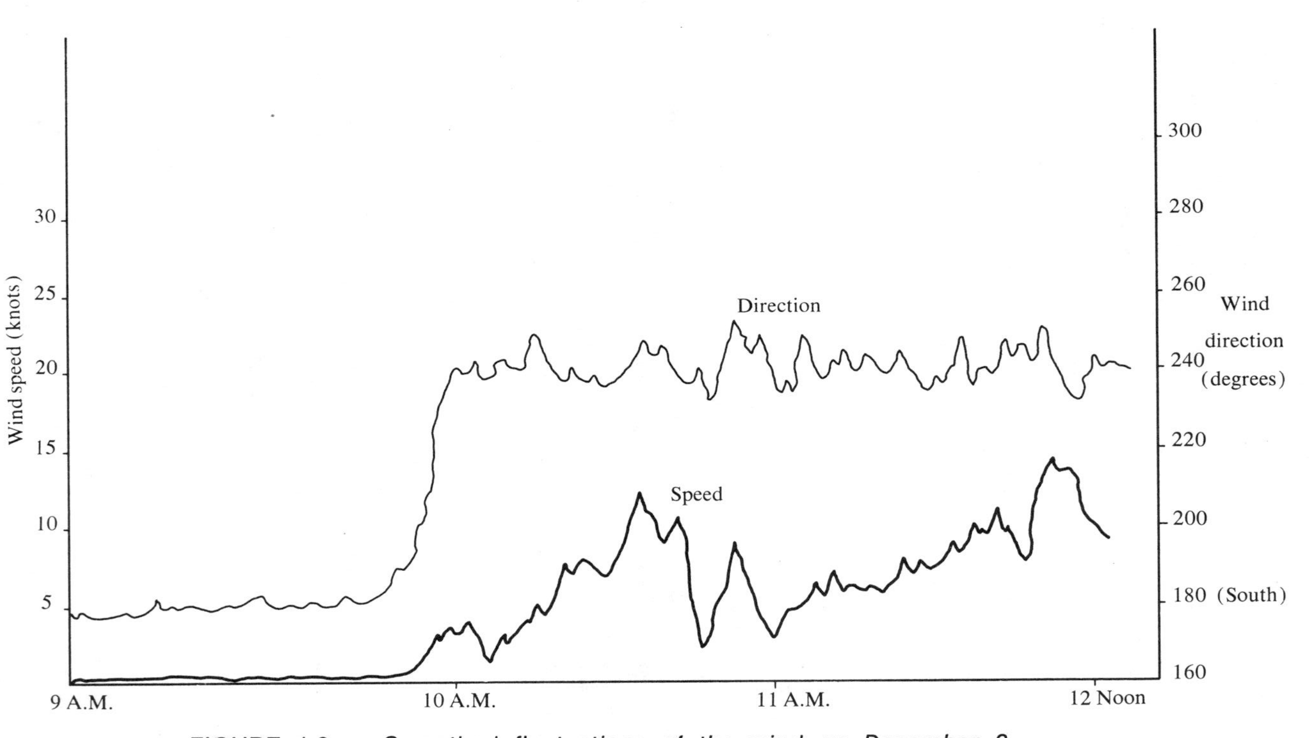

FIGURE 1.6 *Smoothed fluctuations of the wind on December 2, 1978, at State College, Pennsylvania.*

handle and they rarely reached heights greater than 3 kilometers. Airplane soundings were expensive and, at least in the early days, could not provide data to the altitudes desired or during periods of severe weather.

The radiosonde, carried aloft by balloons, transmits its measurements by radio back to a ground station. The radiosonde used in the United States consists of a lightweight, inexpensive radio transmitter that emits a continuous signal. The temperature-measuring and humidity-measuring elements control the frequency or amplitude (intensity) of the audio output of the radio signal. An aneroid barometer cell, moving a contact arm across a series of metal strips, alternately connects temperature and humidity into the circuit. By setting consecutive contacts for known pressure intervals, the temperature and humidity are recorded as a function of pressure. The altitude can be computed if the vertical distribution of temperature, humidity, and pressure is known. The radiosonde is now the principal tool of the meteorologist for systematically observing the conditions of the lowest 30 kilometers of the atmosphere. Exploration of higher levels has been accomplished principally by rockets.

Satellites

Observations of most of the earth's atmosphere are entirely inadequate. About 70 percent of the earth's surface is covered by oceans, and a large proportion of the rest is dominated by mountains, snow, deserts, and jungles. Even in populated areas, the density of weather observing stations permits the construction of only a very "coarse-grained" picture of the atmosphere.

The launching of weather satellites has improved this situation somewhat. Equipped with television cameras, they transmit to earth pictures of the clouds as seen from above. These photographs have been valuable, especially over the oceans, to pinpoint the locations of storms. On a few occasions, hurricanes (Figure 5.17) that were undetected by the low-density oceanic network of stations have been uncovered by weather satellites. Measurements of radiation from the earth also promise to yield increased knowledge of the earth's energy balance (Chapter 3).

There are both polar-orbiting satellites, which view strips of the earth as they complete each orbit, and geostationary satellites, which remain over a fixed spot on the equator. An advantage of the geostationary satellite is that the evolution of weather systems with time can be seen by comparing pictures of the same area at short time intervals (usually ½ hour). Time-lapse movies reveal the spinning cyclones, the growth of thunderstorms, and the translation of all weather systems across the globe.

While the visible satellite photographs reveal much qualitative information about weather systems, satellites are also providing much needed quantitative data, especially over remote ocean areas where conventional observations are scarce. Cloud motions indicate the wind velocity at that level. Infrared pictures yield information on the surface temperature and on the vertical temperature structure of the atmosphere. Sensors that measure microwave energy provide data on the humidity and liquid water content. Although these quantitative data are not yet quite as accurate as rawinsonde data, improvements are being made, and it is

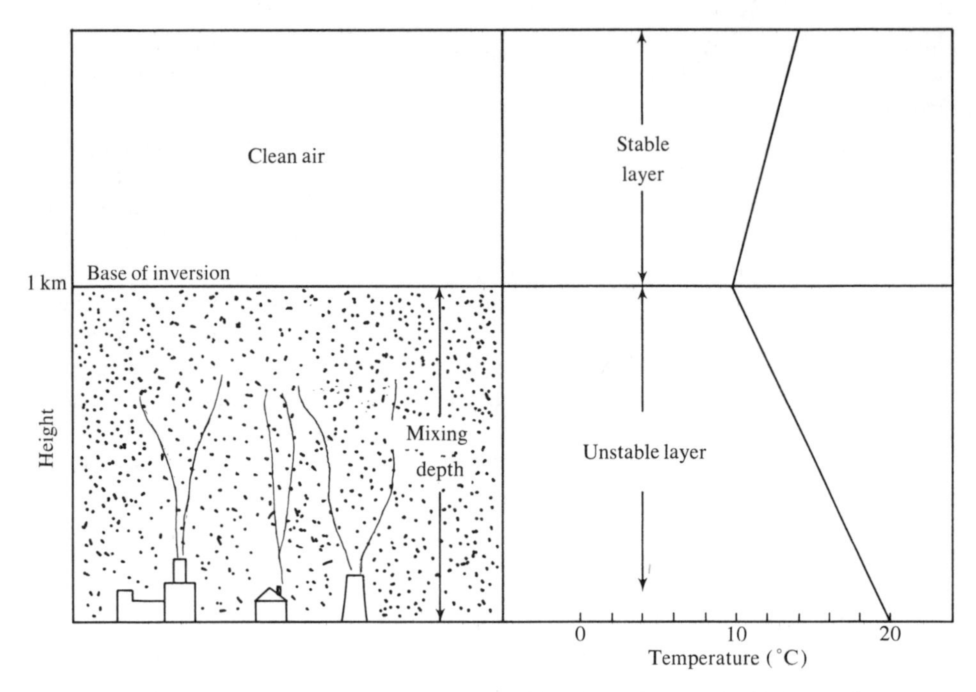

FIGURE 1.7 *Pollutants mix vertically throughout mixed layer because the air is unstable. Above mixed layer, air is stable and vertical mixing is suppressed.*

likely that satellites eventually will provide much of the upper air data that are necessary for analysis and prediction.

1.4 Air Pollution

Many types of air pollution have a variety of adverse effects on animal and plant life. Sulfur dioxide irritates the eyes, nose, and throat, damages the lungs, and aggravates asthma and bronchitis. When dissolved in precipitation, it forms sulfuric acid and can product pH values as low as 3.0, which is acidic enough to dissolve marble statues and some fabrics. Particulates such as dust restrict visibility, dirty surfaces, and reduce the amount of sunlight reaching the ground. Oxidants including nitrogen dioxide and ozone irritate the eyes, aggravate lung diseases, and perhaps accelerate the aging process, as well as cause disintegration of many materials. Carbon monoxide in weak concentrations impairs the oxygen transport function in the body and increases the general mortality rates; in large concentrations it can be fatal. Lead from automobile exhausts accumulates in the body, adversely affecting the brain, kidneys, and nervous system.

Air pollutants may be classified according to whether they are inert or reactive. Inert pollutants are those that react very slowly or not at all with other chemicals or gases in the atmosphere. Inert pollutants are often considered as passive substances because they are transported by the wind and diffused by turbulent eddies without changing their composition. Reactive pollutants, on the other hand, undergo chemical transformation as they encounter other substances or are altered by solar radiation. Pollutants may also be classified as gases or aerosols, which are solid or liquid particles dispersed in the atmosphere. Major gaseous pollutants include ozone, sulfur dioxide, carbon monoxide, and oxides of nitrogen. Aerosols include smoke particles (carbon), bacteria, pollen, salt particles, and many types of dust. When averaged over the entire earth, natural processes account for approximately 90 percent of the total particulate matter suspended in the air.

Concentrations of aerosols are usually expressed as the mass of the pollutant per unit volume of air, e.g., in units of micrograms per cubic meter. The concentration of gases may also be given in terms of mass per unit volume; however, gas concentrations are frequently given in terms of volume concentrations, i.e., the number of pollutant molecules per million molecules of air or parts per million (ppm). The relation between mass and volume concentrations at typical sea-level temperatures and pressures is

$$1\ \mu g/m^3 = \left(\frac{24.5}{M}\right)\ \text{ppm} \qquad (1.4)$$

where M is the molecular weight of the gas.

Aerosols occur over a large range of sizes (Figure 1.1). Their formation occurs in two principal ways, *disintegration* of material and *agglomeration* of molecules. The formation of fine dust particles by the action of wind blowing over dry soil and the production of salt crystals by the breaking of bubbles on the sea surface and subsequent evaporation of the water are examples of the disintegration process. An example of aerosol formation by agglomeration is the production of water drops by condensation. Another example is the formation of soot by carbon molecules that combine after escaping combustion when oil or coal is partially burned.

The chemical composition of aerosols varies widely depending upon nearby sources. Common constituents, with typical concentrations of 10^{-6} grams per cubic meter include sulfate (SO_4), nitrate (NO_3), chloride (Cl), and ammonium (NH_4) ions. Sulfate aerosols with diameters of about 1 μm are the principal ingredients of large-scale pollution episodes which can cover several states and reduce visibilities in a murky haze to several kilometers or less. Sulfate ions are also ingredients of sulfurous smog (the type of smog associated with the infamous "London fogs" which plagued Great Britain from the 13th century until the early 1950s when controls on the burning of sulfurous coal were imposed). On the other hand, photochemical smogs, produced when sunlight causes chemical reactions between oxides of nitrogen and organic compounds, contain large concentrations of nitrate ions.

Aerosols are removed by settling to the surface (*sedimentation*), *adherence* to solid objects after impact, and by wet processes of *rainout* and *washout*. Rainout refers to the incorporation of particles into cloud drops or ice crystals when the particles serve as cloud nuclei (see Section 2.1). Washout is the removal of particles by precipitation falling through polluted air.

Sulfur compounds are among the most prevalent and harmful pollutants. Sulfur dioxide (SO_2) was a major component of the London smog episode of December 1952 which caused an estimated 4000 deaths. Fuels such as oil or coal which contain sulfur produce sulfur dioxide (SO_2) when burned. Copper, zinc, or lead ores often contain sulfur, which is separated from the metal by roasting. Sulfur dioxide may be dissolved in water and oxidized forming sulfuric acid (H_2SO_4), which is the primary cause of acid rain. Because SO_2 can exist for several days, it can be carried by the winds hundreds of kilometers from the source, producing regional air pollution episodes. It can also acidify the rainfall downstream from the source, causing international problems. For example, the Scandinavian countries have accused industries in Great Britain of causing acid rain over Norway and Sweden.

Oxides of nitrogen, consisting of nitric oxide (NO) and nitrogen dioxide (NO_2), are important ingredients in the formation of photochemical smog (the type found in Los Angeles). When heated by combustion engines such as the automobile, atmospheric nitrogen (N_2) is oxidized to NO, which may then be oxidized into the brown gas NO_2. When exposed to the short wavelengths of sunlight, the NO_2 is broken up into NO and free oxygen O. This atomic oxygen then combines with molecular oxygen O_2 to form ozone O_3, an irritating gas that causes rapid disintegration of materials such as rubber or plastic. Many additional reactions can occur between the ozone, oxides of nitrogen, and reactive hydrocarbons to produce the complicated mixture of photochemical smog.

Effect of meteorology on air pollution

The two most important aspects of meteorology that affect air pollutant concentrations are *wind speed* and *stability*. For a single source of a given emission rate, the volume of air affected by the source increases linearly with the wind speed, hence, other factors being equal, the concentration from a single source varies inversely as the wind speed. The stability of the atmosphere also plays an important role by governing the intensity of turbulence which mixes air pollutants vertically and hence dilutes the polluted mixture. Under *unstable* conditions, with warm air near the ground and colder air aloft (as usually occurs during sunny days), vertical mixing occurs readily, with dirty air carried aloft and cleaner air mixed downward. Under *stable* conditions, when the temperature decreases slowly with height or even increases with height (an *inversion*), vertical mixing is inhibited and pollutants tend to remain at a constant elevation in the atmosphere.

Under normal conditions, a layer of unstable air occurs next to the ground with stable air aloft. The depth of this unstable layer in contact with the ground is called the *mixing depth* (Figure 1.7), because vertical mixing occurs easily in this layer. Typical depths of the mixed layer vary from about 600 m over Los Angeles to a kilometer or two over cities in the Midwest or East. Pollutant concentrations tend to be vertically constant in the mixed layer. Therefore, the greatest pollution concentrations occur with shallow mixed layers, or when an inversion occurs at the surface and vertical mixing is weak everywhere.

The effects of stability and wind speed form the basis for a simple prediction model of air pollutant concentration downwind of a single source:

$$X = \frac{CQ}{U} \quad (1.5)$$

where X is the concentration in micrograms per cubic meter, Q is the source rate in micrograms per square meter per second, U is the wind speed, and C is a dimensionless number that depends on the stability. Values of C range from 600 for stable to 50 for unstable conditions.

PROBLEMS

1. If the average number of molecules per cubic centimeter is 25×10^{18} and the average density is 1.2×10^{-3} g cm^{-3} at sea level, what is the average mass of an air molecule?
2. Modern jet aircraft often fly at an elevation of 10 km at which level the pressure is about 300 mb. If the temperature is $-40°C$ at this level, what is the density? (Express the answer as a fraction of the average surface density given above.)
3. Typical stratus clouds occur at an elevation of 1 km above the ground and have droplets of diameter 10^{-3} cm. How long would it take such a drop to reach the ground if there were no vertical air motions?
4. The average sea-level pressure is 1013 mb. What is this average expressed in inches of mercury?
5. If the global average sea-level pressure is 1013 mb and the average temperature is 10°C (283 K), what is the average density at sea level?
6. If you were to construct a scale model of the earth and its atmosphere, starting with a 1-meter-diameter globe, how far from the surface would the following extend?
   (a) Mt. Everest,
   (b) the level at which 99 percent of the atmosphere is found,
   (c) the level of the tropopause.
7. Compute the height of a water barometer at a place where the atmosphere's pressure is 850 millibars, if the temperature of the barometer is 10°C.
8. Make a list of at least a dozen ways in which the atmosphere—its constituents and its motions—affect people and their activities.
9. As far as "weather" is concerned, which of the atmosphere's gases are most important? What role(s) does each play in the weather-making processes?
10. The average atmospheric density at sea level is about 1.2×10^{-3} g/cm^3, and the average pressure is 1013 millibars. If the density were constant in the vertical, what would be the depth of the atmosphere (at what point would $p = 0$)? What would be the depth of liquid water having the same pressure (1013 millibars) at the bottom?
11. If there is a fog in the morning with a temperature of 0°C, what will be the relative humidity in the midafternoon, when the temperature is 10°C, assuming that the actual vapor pressure remains unchanged?

12. In one of the worst pollution episodes in the United States (October 26–31, 1948), nearly half the population of Donora, Pennsylvania became ill and approximately 20 people died from the high concentrations of pollutants. Particulate concentrations reached 4 mg/m^3 while SO_2 concentration exceeded 0.5 ppm. What was the SO_2 mass concentration in micrograms per cubic meter?

13. A moderate SO_2 emission rate is 1 $g/km^2/s$. Estimate the concentration of SO_2 in micrograms per cubic meter for the following conditions and compare the values with the peak SO_2 concentrations in the Donora, PA episode:
    (a) Unstable, wind speed = 5 m/s.
    (b) Stable, wind speed = 1 m/s.

Clouds and Precipitation

2.1 Clouds

In the discussion on humidity, it was implied that as soon as the concentration of vapor begins to exceed the saturation value, which is largely dependent on temperature, the excess moisture becomes liquid or solid. It is not quite that simple. There are surface intermolecular binding forces at the boundary of a liquid or solid that restrain the energetic molecules from escaping. (These binding forces produce what we commonly refer to as the "surface tension" of a liquid—the force that resists rupture of the surface, which we feel when we dive into a pool of water.) Whether or not a molecule will escape this surface restraining force depends on the molecule's speed when it reaches the surface. Although the *average* speed of molecules moving in a random fashion depends on the temperature, *individual* molecular speeds vary over a broad range of values. Some molecules may have sufficient momentum in the direction normal to the surface to escape; the higher the water temperature, the greater the percentage of molecules that will have the critical speed.

Even after some molecules have escaped from the bulk water, they continue to move about in all directions and over a range of speeds, and inevitably, some may strike the surface, penetrate it, and be recaptured. In a given time interval, when more molecules are escaping than are being captured, we say there is *evaporation;* when the converse is true, there is *condensation;* and when the escape and capture rates are equal, there is an *equilibrium* state. The molecules of water vapor above the surface exert a pressure that depends on their number per unit volume and their temperature. Pressure exerted in the equilibrium state, with as many molecules condensing on the water surface as are evaporating from it, is called the *saturation vapor pressure.* The term *transpiration* is commonly used to describe evaporation of moisture from leaves and other plant surfaces.

The equilibrium at saturation between evaporating and condensing molecules is altered if either the water surface is curved (because of surface tension effects) or if the water has impurities dissolved in it. Therefore, each water droplet has an equilibrium relative humidity slightly different from 100 percent. At environmental humidities greater than this equilibrium value the drop will grow; at humidities less than this value the drop will evaporate.

For condensation, sublimation, or freezing to occur there must be a suitable surface available. Dew or frost forms easily on grass, soil, windows, etc., whenever the air temperature reaches the dew point or the frost point. But in the free atmosphere there are no such extensive surfaces.

In pure air (i.e., all gas with no "foreign" particles, water droplets, or ice crystals) condensation or sublimation is extremely difficult to achieve, even under highly super-saturated conditions (relative humidity much greater than 100 percent). For a single droplet to form, it would be necessary for many water molecules in the air not only to collide with one another but also to stick together. However, the probability of such multiple collisions is extremely small and even when some do stick, thermal agitation of the molecules tends to cause some of the outer molecules to escape from any small "embryo" that may form. Only after a droplet has reached a critical size (radius about 10^{-5} cm) are the binding forces sufficient to hold more of the molecules that strike its outer surface than the number that escape it. Since molecular speed depends on temperature, this critical size is a function of temperature. Once this critical size is reached, further growth is very possible, but the probability that the hundreds of millions of molecules required to produce such a size will not only collide but also stick together is extremely small in pure air.

Fortunately, in the natural atmosphere there are numerous particles much larger than individual molecules that provide surfaces or condensation nuclei to which water molecules can adhere. It is around these that water droplets and ice crystals grow. (Cloud drops and raindrops are never "pure," despite the myth that they are, although the proportion of "foreign" particles to water is usually very small.) Certain types of particles, such as salts injected into the atmosphere from the sea, attract water molecules to their surfaces and are said to be hygroscopic nuclei and the effect of hygroscopic nuclei on the equilibrium relative humidity is called the solute effect. On these, condensation may actually begin well before the air becomes saturated. However, salt nuclei represent only a small number of the total particles suspended in the air, and there are a large number of other types of particles, such as combustion products, meteoritic dust, and soil, that also serve as nuclei. These small particles, most of which have diameters of less than 1 micrometer (a thousandth of a millimeter) and are found in quantities of 10,000 or more per cubic centimeter, are so small that they remain suspended in the air for days at a time. (See Figure 1.1.)

While hygroscopic particles allow water vapor to condense at relative humidities below 100 percent, the curvature of the tiny spherical droplets produces an effect that requires higher relative humidities for condensation than would be required for a flat surface of water. Thus a drop of pure water with a radius of 0.1 μm requires a relative humidity of 101 percent for equilibrium; at a humidity of 100 percent this drop would evaporate.

At temperatures below freezing, when the saturation vapor pressure is approached, water vapor molecules may be converted directly to ice crystals, a process called sublimation. As with condensation, sublimation requires special tiny particles. The particles, called *sublimation nuclei,* are very rare in the atmosphere, and hence nearly all clouds consist at first entirely of water drops and are produced by condensation rather than sublimation. Liquid water drops at temperatures below freezing are said to be supercooled.

Pure liquid water will not freeze without special nuclei called freezing nuclei. There is evidence that certain materials are more effective as freezing nuclei at warmer temperatures than others. In fairly large volumes of water, the likelihood of having at least a few effective nuclei is quite high and freezing normally occurs at a temperature very close to 0°C. But in very small droplets, it is possible to achieve a temperature as low as −40°C before freezing takes place. Indeed, in the atmosphere it is common for liquid water drops to exist in clouds at temperatures as low as −20°C. Large drops freeze at a higher temperature than small drops.

Fog and clouds are composed of liquid water and/or ice particles suspended in the air. There can be as many as 500–600 particles in each cubic centimeter, although normally there are fewer than half this number. However, the particles in nonprecipitating clouds are normally quite small (averaging about 0.01 millimeter in radius and rarely exceeding 0.1 millimeter), and so they fall to earth very slowly. In calm air, a droplet having a radius of 0.05 millimeter falls at the rate of less than ½ m/s (1 mi/h). This maximum fall velocity, known as the terminal velocity (Figure 1.1), is imposed by air resistance. Most clouds are formed when air is rising; so in practice, even in extremely weak upward air currents, drops of this size can be suspended in the atmosphere for many hours. In fact, clouds begin to precipitate only when some of the drops within them reach sufficient size to fall through the air with an appreciable velocity.

Cloud types

Aside from observation of the internal makeup of clouds, the outward appearance of clouds is of significance to the meteorologist in interpreting the physical processes in the atmosphere, and it is often a harbinger of the weather to come. The basic clouds are identified on the basis of their form and the approximate height above the ground where they normally occur. The names of the basic clouds are composed of the following roots: *cirrus*, (feathery or fibrous); *stratus* (stratified or in layers); *cumulus* (heaped up); *alto* (middle); and *nimbus* (rain). The ten basic clouds are:

High (base over 7 km, or 23,000 ft, generally composed entirely of ice crystals): *cirrus* (Ci), *cirrostratus* (Cs), *cirrocumulus* (Cc)
Middle (2–7) km, or 6500–23,000 ft): *altocumulus* (Ac), *altostratus* (As)
Low (below 2 km, or 6500 ft): *stratus* (St), *stratocumulus* (Sc), *nimbostratus* (Ns)
Clouds of vertical development (base usually below 2 km, or 6500 ft, but top can extend to great heights): *cumulus* (Cu), *cumulonimbus* (Cb)

Illustrations of these clouds appear in U.S. Weather Bureau publications, encyclopedias, and many other places and so will not be given here. Some common adjectives applied to the basic names to further describe particular clouds are:

Uncinus Hook-shaped—applied to cirrus, often shaped like a comma
Castellanus Turreted—applied most often to cirrocumulus and altocumulus

Lenticularis Lens-shaped—applied mostly to cirrostratus, altocumulus, and stratocumulus; occurs where air currents are undulating sharply in the vertical, as sometimes occurs on the lee side of mountains
Fractus Broken—applied only to stratus and cumulus
Humilis Lowly—poorly developed in the vertical; applied to cumulus
Congestus Crowded together in heaps, like a cauliflower —applied to cumulus

As will be shown in the next section, almost all clouds result from the rapid cooling of air when it ascends. In stratiform clouds, the motion in the vertical is generally small (less than 20 cm/s), while in cumuliform clouds, the upward and downward velocities are much stronger (up to 30 m/s or more).

Fog is merely a cloud in which the observer is immersed. It is thus a suspension of small water droplets. When the fog is formed of tiny ice crystals, it is known as *ice fog*. Haze often precedes and follows fog. Haze is formed of very small droplets of wet hygroscopic particles. The air does not feel "wet" as it does in fog.

Smog originally meant a natural fog contaminated with pollution, i.e., a mixture of smoke and fog. It has recently come to mean any general urban air pollution mixture in which visibility is significantly reduced, whether or not natural fog is actually present. Therefore, there are a variety of smogs, some produced through photochemical reactions between nitrogen oxides and hydrocarbons (both emitted primarily by automobile exhausts) and some produced by sulfurous fumes emitted by combustion of fuel oils and coal of high sulfur content. Because the chemical constituents of smogs are often hygroscopic, tiny water drops can exist at relative humidities as low as 80 percent. Actually, these drops are often drops of sulfuric acid, with pH values as low as 3.0.

Formation of different cloud types

Because of the abundance of cloud nuclei in the atmosphere, clouds will be formed whenever the actual vapor pressure exceeds the saturation value. Theoretically, this could occur by either increasing the actual vapor pressure or decreasing the saturation vapor pressure. Since it is impossible to increase the actual vapor pressure past the saturation value by evaporation, the only alternative is to decrease the saturation value. Such a decrease is accomplished when the air temperature is lowered, because the saturation vapor pressure decreases rapidly with decreasing temperature (Figure 1.5). The required drop in temperature can occur through mixing with colder air or by cooling of the entire air parcel.

The formation of clouds by mixing is illustrated in Figure 2.1 which shows a graph of the saturation mixing ratio over water at a pressure of 1000 mb. The mixing ratio is defined as the mass of water vapor per mass of dry air; the saturation mixing ratio is the value at saturation. Suppose there are two parcels of air at different temperature and mixing ratios, as indicated by points A and B on Figure 2.1. Because the mixing ratio of both parcels is less than the saturation value, clouds will not exist in either parcel. Consider what happens if the two parcels are mixed, however. For parcels of equal mass, the temperature and mixing ratio of

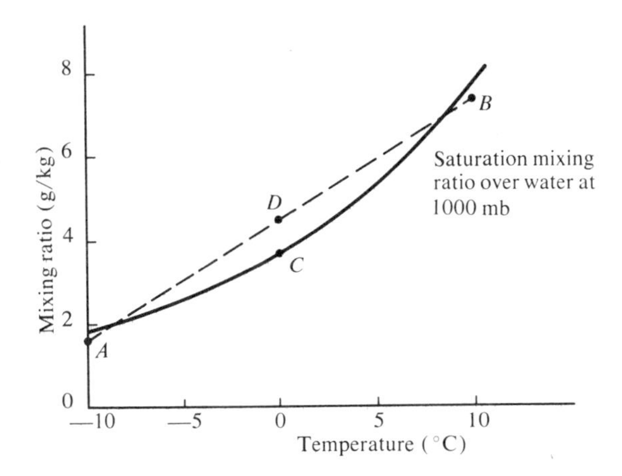

FIGURE 2.1 *Formation of a cloud by mixing parcels A and B.*

the mixture will be the average of the original values in each parcel. This average of 0°C and 4.6 g/kg is shown by point *D* in Figure 2.1. Note that because the saturation mixing ratio curve is concave, the mixing ratio of the mixture exceeds the saturation mixing ratio at the temperature of the mixture. Thus a mixing of two unsaturated parcels of air has produced a parcel that is supersaturated, and one in which cloud formation would be very likely. An example of cloud formation by mixing is sea smoke (or steam fog), which is formed when cold air flows over warm water and mixes with the warm, moist air in contact with the water. Another example is the condensation trail (contrail) produced when hot, moist exhaust from jet engines mixes with cold environmental air. But perhaps the most familiar example of a cloud produced by mixing is the cloud formed when moist breath is exhaled on a cold day.

The cooling of extensive layers of air to the dewpoint is the most common mechanism for producing clouds. Cooling may occur by radiation, by conduction when air comes into contact with a colder surface, or through lifting. Radiative and conductive cooling produces fog or stratus clouds. Because radiation and conduction are slow processes, these clouds are not very thick and usually do not produce any precipitation. The mechanism that produces the most rapid rate of cooling and is, therefore, responsible for almost all precipitation is the cooling of rising air by expansion. As discussed in Section 4.3, dry air cools 9.8°C/km of rise. With condensation and the accompanying release of latent heat, the rate of cooling is less but still exceeds 6°C/km. Therefore, a parcel of air in a thunderstorm updraft can cool 40°C in 15 min as it is lifted 8 km or more. Even the large, nonconvective cloud sheets associated with middle-latitude storm systems can ascend over 4 km in one day, producing cooling of 30 to 40°C during this time. Vast sheets of cirrostratus, altostratus, and nimbostratus are produced in the region around cyclones (low pressure systems) in which the air is moving upward.

When the air is stable (see Section 4.4), small-scale vertical motions are suppressed and clouds take on a smooth, uniform, layered appearance. When the atmosphere is less stable, small-scale, saturated parcels of air may become buoyant.

As many of these buoyant elements rise, a turreted appearance to the clouds evolves. Thus *castellanus* clouds indicate relatively unstable layers of air.

Waves of alternating upward and downward motion often exist in the atmosphere. If the air is close to saturation, the small amount of lifting can produce a limited region of condensation, and a wave cloud results (see Figure 2.2). Lifting of air over mountains often produces a similar *lenticularis* cloud.

Evaporation

So far, we have considered only the transition of water from the gaseous to the liquid or solid states and the special conditions required for the formation of drops and ice crystals in the atmosphere. We next consider the transition rate in the other direction; i.e., evaporation (from liquid to gas) and sublimation (from solid directly to gas).

If the volume above a wet surface is not already saturated, then the gaseous water molecules will spread into it (diffuse) until the entire volume available becomes saturated with water molecules. Thus, evaporation proceeds as long as the vapor pressure in the free air is less than the saturation value at the surface. If the air temperature is much lower than the surface temperature, evaporation continues as long as the vapor pressure in the air is lower than that at the surface, but as soon as the air reaches its saturation value, the excess vapor will be condensed.

The factors that affect the rate of evaporation from a surface are: (1) the temperature of the water surface (since high temperature implies a high saturation vapor pressure); (2) the actual and saturation vapor pressures in the free air near the surface; and (3) the effectiveness of diffusion of the vapor away from the surface. As we saw in our discussion of mixing by the wind, the diffusion of water vapor is greatest when the wind is strong and gusty.

Wind has another effect that enhances the evaporation over large water bodies. Spray ejected into the air from wind-generated waves is composed of very small droplets (diameter less than 0.05 millimeter), which evaporate rapidly in the air. When the wind speed is greater than 50 km/h (35 mi/h), the rate of evaporation of spray may actually exceed the rate from the surface of the water body.

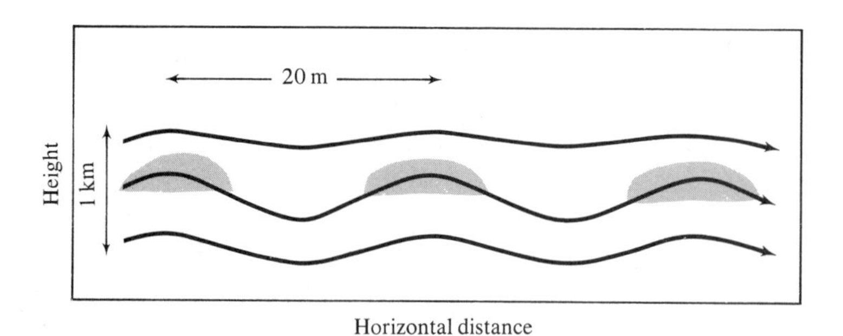

FIGURE 2.2 *Wave clouds formed when air in the rising portion of the wave is cooled below the dewpoint.*

Direct measurement of the worldwide evaporation rate is not practical, and estimates are based on computations from those theoretical relationships given above. Evaporation pans (shallow, circular pans filled with water with a scale to indicate the changes in depth) are often used to "measure" the evaporation from dams and reservoirs and over cultivated fields. However, the evaporative loss from such pans is almost invariably higher than the true evaporation because of differences between pan and natural surface properties and the difference in "fetch," (i.e., the distance travelled by the air in traversing the surface).

Most of the world's evaporation occurs over the oceans, and it is especially high throughout the year over the tropical latitudes between 5 degrees and 30 degrees (4–7 millimeters per day) and over the western halves of oceans in middle latitudes during the cooler half of the year (4–9 millimeters per day) when dry, cool air from the continents sweeps over such warm ocean currents as the Kuroshio and the Gulf Stream.

2.2 Precipitation

Raindrops are usually between one and several millimeters in diameter. Since the average drop diameter in a nonprecipitating cloud is about 0.02 millimeter, this means that many cloud drops have increased their volume by a factor of 1,000,000 by the time they fall out of a cloud. Such growth cannot be explained merely by further condensation of water vapor on existing water particles since the process becomes very slow for larger drops. Small particles must therefore unite to form large particles. How this is accomplished is still not completely settled, but the most probable mechanisms are the following:

(1) *Collision and coalescence of particles.* The drops formed in a cloud are not all of the same size. Due to differences in the rate at which condensation proceeds in different parts of a cloud—sometimes separated by very small distances—the largest drops in the cloud may have diameters several times greater than the smallest. As the air swirls about, the larger drops, because of their greater mass, have more inertia than the smaller ones. These large drops tend, therefore, not to follow exactly the same path as the small ones and, as a result, collide and often coalesce (combine) with the small ones. Repeated collision and coalescence by a drop may cause it to grow so large that it splinters into several drops, which in turn grow by collision and coalescence, thus producing a "chain reaction" of raindrop growth.

(2) *Growth of ice crystals.* Frequently, especially in the middle latitudes, the uppermost layers of clouds will be composed of ice crystals, while the lower layers contain supercooled (at temperatures below 0°C) liquid drops. Through stirring within the cloud, or because of different fall velocities of the particles, ice crystals and supercooled drops become mixed. At the same temperature, the saturation vapor pressure over a liquid surface is greater than that over an ice surface (see insert of Figure 1.5), and the ice crystals will therefore grow at the expense of the water drops. This growth mechanism, which is called the *Bergeron process* after the Swedish meteorologist who first suggested it, is believed to be quite important in the

initiation of precipitation, although further growth most likely involves the collision-coalescence of particles described in the previous paragraph.

The importance of the coexistence of ice crystals and undercooled water drops in the initiation of precipitation forms the basis for many modern cloud-seeding experiments. In a cloud that contains few or no ice crystals, dry ice introduced into the cloud may cool enough drops to their freezing point to produce crystals for later growth by the Bergeron process. Undercooled droplets can also be induced to freeze through injection of certain types of materials, such as silver iodide, apparently because the crystal structure of these materials is very similar to that of ice (Figure 2.3).

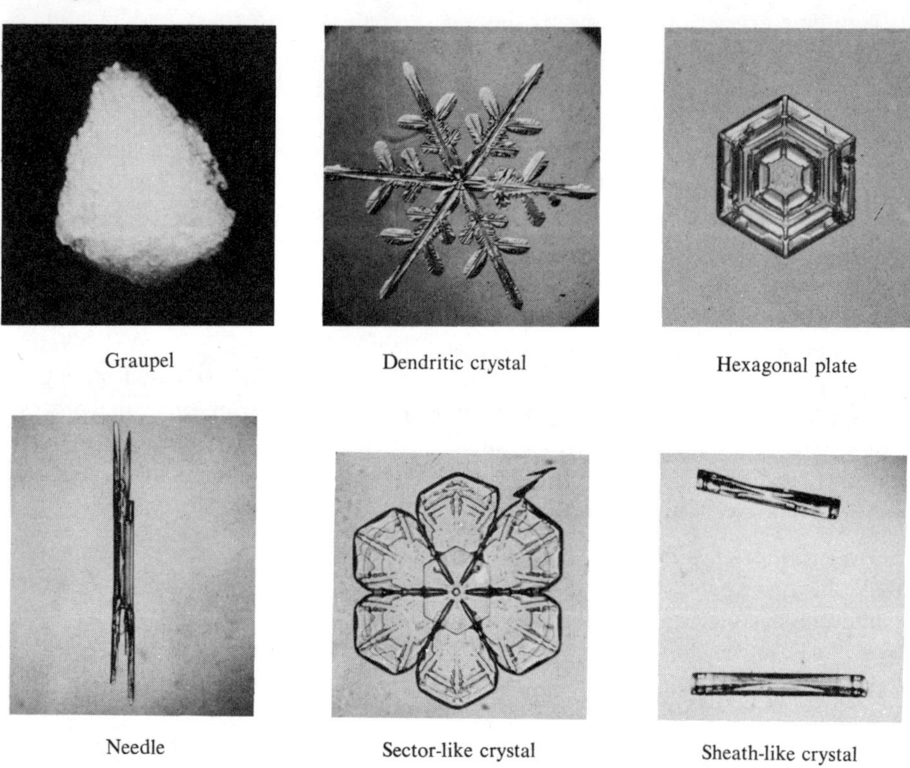

FIGURE 2.3 *Some forms of snow crystals. (Courtesy of C. Magono, Hokkaido University, Japan.)*

Precipitation types

The only difference between *drizzle* and *rain* is the size of the water droplets. The diameter of the former is generally less than 0.5 millimeter. The principal solid forms of precipitation are:

Freezing rain or drizzle Rain or drizzle that freezes on impact with the ground or objects

Sleet Small ice particles or pellets that originated as rain but froze as they traversed a cold air layer near the ground

Hail Small balls or chunks of ice with a diameter of 5–50 millimeters (0.2 – 2 inches) or more that fall from cumulonimbus clouds. These destructive stones are formed by the successive accretion of water drops around a small kernel of ice falling through a thick cloud; as each drop is frozen onto the nucleus, it may form a new shell, so that many hailstones acquire an onion-like cross section (Figure 2.4).

(a)

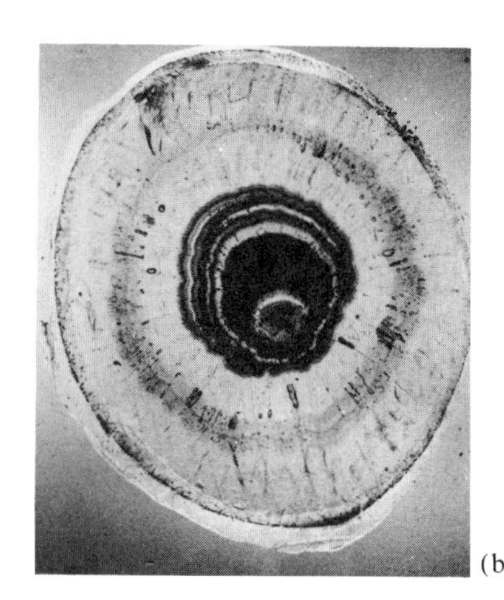
(b)

FIGURE 2.4 *(a) Hailstones. (b) Cross section of a hailstone. (Courtesy of R. List, SLF.)*

Measurement of precipitation

For practical purposes of water supply, meteorologists are concerned with measuring the amount of water reaching the earth's surface. This is done by sampling the depth of water that would cover the surface if the water did not run off or filter into the soil. A rain gauge is merely a collection pail with a ruler to measure the depth of water. The depth is usually measured in increments of a hundredth of an inch or a millimeter. Solid forms of precipitation are melted and the equivalent liquid depth recorded. In the case of snow, which may remain on the ground for a long period of time and thus serve as a natural water reservoir, the snow depth is of interest. Normally, the ratio of snow depth to liquid equivalent is about 10 to 1, although sometimes it is as much as 30 to 1.

Precipitation amounts vary considerably from place to place, even during a single storm, so that the problem of adequate sampling over an area is a serious one. In most places in the world, not more than one 20-cm diameter rain gauge is installed in every 200 square km, which is a ratio of areas of 314 cm^2 to $(200 \times 10^5)^2$ cm^2, or approximately $1:10^{12}$. This is somewhat like taking a single hair from one Californian's head to characterize the hair of everyone in the state. In mountainous areas especially, amounts may vary by a factor of two or three in a distance of less than 20 km. (Along the northeast slopes of Hawaii, the annual rainfall varies from 40 cm to over 750 cm in a distance of about 25 km.) Interpretation of measured amounts must be done with considerable care. Even at the same point, annual amounts vary greatly, especially in semi-arid climates; the year-to-year amounts can easily fluctuate by 50 percent or more of the long-term average annual precipitation. For this reason, claims by rainmakers that they have increased the rainfall by some precise figure, like 10.4 percent, should be viewed with a great deal of skepticism.

The hydrologic cycle

Humans have always been highly dependent on water supply. Ancient civilizations have withered or prospered as precipitation patterns have shifted over the centuries. Today, the increasing growth and concentration of population and of industry are increasing the demands on an essentially fixed world water supply, and we have become more concerned than ever with the need to carefully budget our water resources.

One of the hydrologist's tasks is to study the distribution of the world's water supply. The earth's total water supply (about 1.46×10^9 km^3) is contained mostly in the oceans (96 percent) and in snow and ice fields (2 percent). Only about 2 percent is contained in the soil, underground water, rivers, and lakes of continents. Even this relatively small proportion of water over the continents would eventually drain into the oceans if there were no transport of water from the oceans to the continents. The atmosphere, although containing only about 0.001 percent of the total global water on the average, provides the vital "hydrologic link" between oceans and land.

Even if all of the water vapor carried in the atmosphere at any moment could be precipitated, an average depth of only about 2.5 centimeters of liquid water would result. But new water is constantly being added through evaporation [at the rate of 5.1×10^{17} liters per year (1.4×10^{17} gal/yr), 85 percent coming from the seas], as well as being extracted through precipitation. Within the moving currents of air, the amount of water passing over an area during a period of time can be considerable. For example, even over arid Arizona, the atmosphere carries as much water in a single week in July as flows in the Colorado River during an entire year. Actually, even during heavy precipitation, only a small percentage of the total water in a vertical air column is precipitated; the intense inflow of air into a storm (as in a hurricane) more than compensates for the low rate of moisture extraction. On the average, only about one-tenth of the total water vapor that passes over the U.S. actually precipitates.

Although the worldwide precipitation rate of 100 cm/yr must equal the global evaporation rate, precipitation and evaporation are far from uniform throughout the world. Of the total water supply from precipitation, 78 percent falls over the oceans. The average annual amounts falling on continents vary enormously—from zero centimeters to more than 700 centimeters (276 inches), with some points receiving as much as 1100 centimeters. In the conterminous United States, the average annual rainfall is about 75 centimeters (30 inches), but ranges from as little as 5 centimeters (2 inches) in the deserts of the Southwest to almost 380 centimeters (150 inches) in the Cascade Mountains of Washington and Oregon.

Of the precipitation that falls over continents, about 65 percent is returned to the atmosphere through evaporation from soil, lakes, rivers, animals, and vegetation; the rest runs off into rivers (some via underground streams), eventually emptying into the oceans. The average evaporation rate from oceans is almost double that over land (540 mm/yr), and since the area of the oceans is almost triple that of land, the oceans account for about 85 percent of the worldwide evaporation.

Hydrology deals with all factors that affect water supply over land—precipitation, storage in soil and underground reservoirs, runoff, evaporation, and the return of the moisture to the atmosphere and oceans. In the United States, a network of more than 13,000 precipitation gauges measures rain and snow; more than 300 evaporation pans are used to estimate evaporation. Storage of water in snowfields is obtained by measurements of snow depth. The flow rate and depth of rivers are also measured. From such information, the hydrologist attempts to anticipate destructive floods, drought, and future water supply, and to design reservoirs, sewer systems, bridges, etc.

Because snowfall builds up over the winter season and is released relatively rapidly when spring arrives, flooding may result. Also, hydrologists need to estimate the runoff available for agricultural purposes. Recently, satellites have been used to estimate snowpack over large unaccessible regions, such as the mountainous areas of the western United States. Figure 2.5(a) shows NASA's Landsat satellite's view of the snow cover over the Sierra Nevada in February of a normal year (1975). Figure 2.5(b) shows the same area during the drought year of 1976–1977.

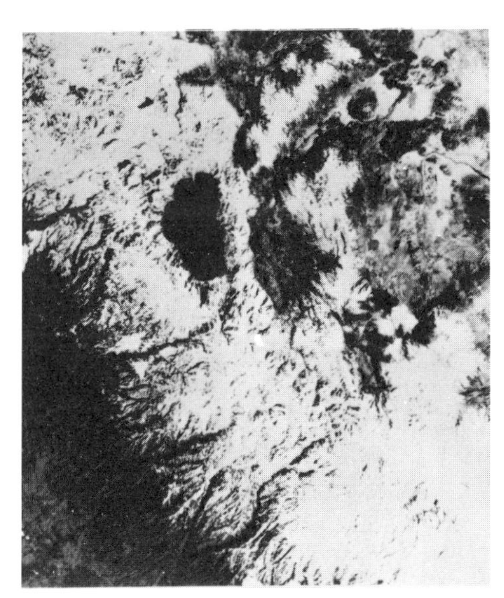

FIGURE 2.5 *Landsat views of snow cover on the Sierra Nevadas during (a) a normal runoff season (1975) and (b) a drought year (1977). (NASA photograph.)*

PROBLEMS

1. The evaporation-precipitation cycle behaves like an enormous desalinization plant. If we wanted to double the natural desalinization rate, how much energy would be required each year? How does this compare with the U.S. power production?
2. How "pure" is the typical raindrop; i.e., what is the ratio of solute mass (condensation nucleous) to water mass? (Assume that the typical nucleus has a diameter of 10^{-4} millimeters and a density 2.5 times that of water and consult Figure 1.1 for average size of raindrop.)
3. The atmosphere contains about 2.5 centimeters of precipitable water and precipitates an average of 100 cm/yr. What must be the mean "residence time" of water in the atmosphere?
4. A typical cloud water content is 1 g/m^3. If a cloud with this amount of water were 3 km thick and all of the cloud water were precipitated out, how deep would the precipitation be?
5. Explain why there is some truth to the proverb "It's too cold to snow."

The Atmosphere's Energy

3.1 Energy Budgets and Heat Engines

A convenient way to examine the workings of the atmosphere is through an energy budget. The law of the conservation of energy requires that we account for all of the energy received by the earth, so by looking at all forms of energy and transformations we have a guide to atmospheric phenomena. This is similar to following in detail what happens to the fuel energy provided in an engine, thereby ending up with a fairly good picture of the operation of the engine. Figure 3.1 presents a schematic energy flow diagram. This and the following chapters will deal with particular portions of the flow diagram.

Practically all (99.98%) of the energy that reaches the earth comes from the *sun*. Intercepted first by the atmosphere, a small part is directly absorbed, particularly by certain gases such as ozone and water vapor. Some of the energy is reflected back to space by the atmosphere, its clouds, and the earth's surface. Some of the sun's radiant energy is absorbed by the earth's surface. Transfers of energy between the earth's surface and the atmosphere occur in a variety of ways, such as radiation, conduction, evaporation, and convection. Kinetic energy (energy of air in motion or wind energy) results from differences in temperature within the atmosphere in much the same way that a heat engine converts differences in heat levels between the inside and outside of the expansion chamber to the motion of the piston. And, finally, turbulence and friction are constantly bleeding off some of the energy of motion, converting it to heat. The combination of these many processes, which are listed in Figure 3.1, produces the complex atmospheric phenomena that determine our weather.

Transfer of heat energy

Heat energy can be transmitted from one place to another by conduction, convection, and radiation.

Conduction is the process by which heat energy is transmitted through a substance by point-to-point contact of neighboring molecules, even though the molecules do not leave their mean positions. Solid substances, especially metals, are usually good conductors of heat, but fluids such as air and water are relatively poor conductors. Heat conduction in air is so slow that it is of little importance

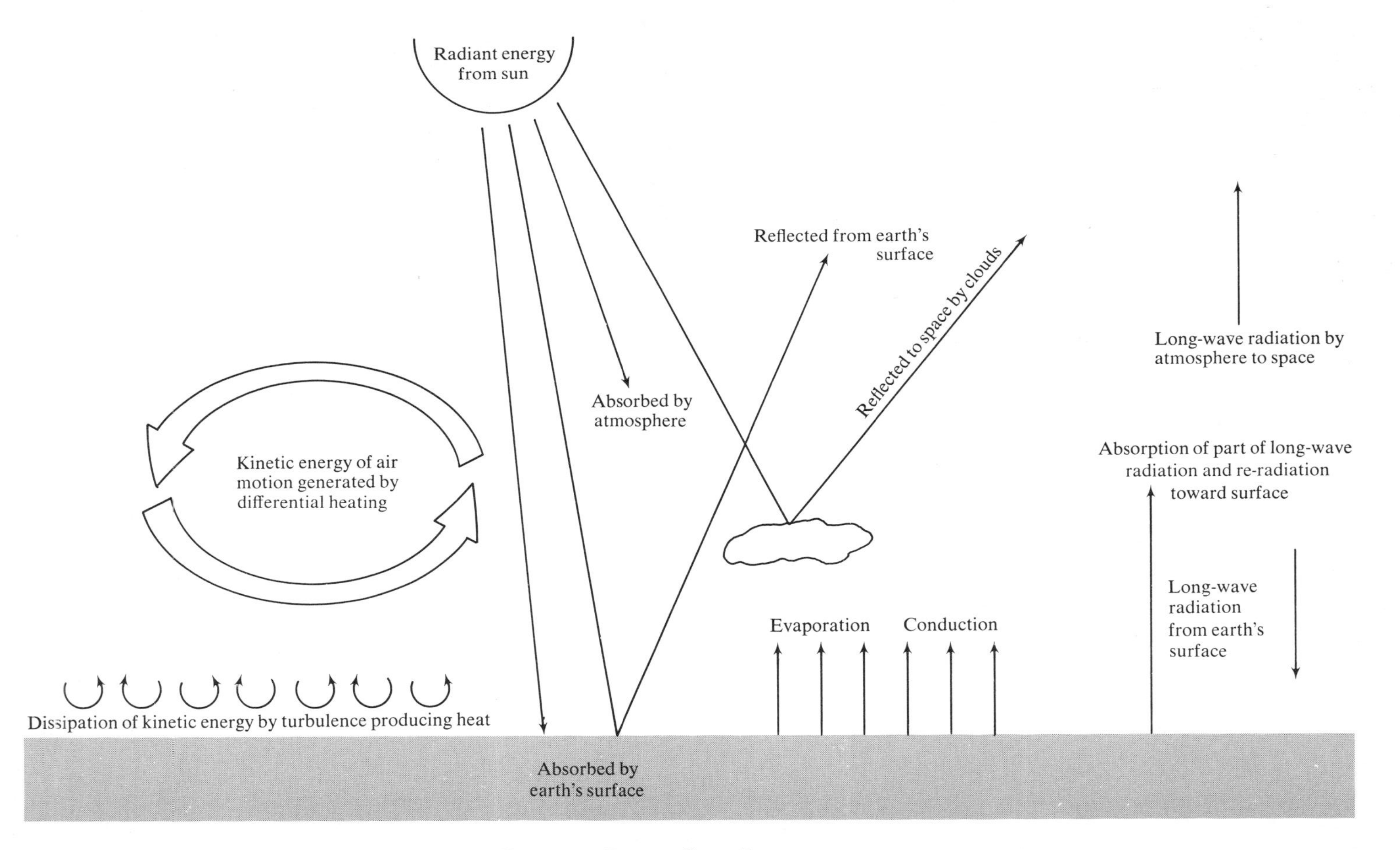

FIGURE 3.1 *Energy flow diagram.*

in transmitting heat within the air itself; however, it is significant in the exchange of heat between the earth's surface and the air in contact with it.

Convection transmits heat by transporting groups of molecules from place to place within a substance. Thus, convection occurs in substances in which the molecules are free to move about; i.e., in fluids. The convective motions that carry heat from one point to another within a fluid arise because the warmer portions of the fluid are less dense (recall the equation of state, p. 10) than the surroundings and therefore rise, while the cooler portions are more dense and therefore sink. A circulation of fluid, i.e., a closed circuit of fluid motion, is thus established between the warm and cool regions. Much more will be said about convection in later chapters because it is an important heat transfer process in the atmosphere above the lowest half-meter or so.

Radiation is the transfer of heat energy without the involvement of a physical substance in the transmission. Heat may therefore be transmitted through a vacuum, and if the radiation takes place through a completely transparent medium, the medium itself is not heated or otherwise affected. The transfer of the heat energy from the sun to the earth is by means of the radiative process. Earth also loses its heat energy to outer space in the same way: by radiative transfer.

3.2 Solar Energy

Since the atmospheric "heat engine" is powered by solar energy, we start our discussion with the source of the "fuel," the sun. The sun is not an unusual star, either in brilliance or in size. A slowly rotating body of hot (several million degrees Celsius), very dense gas, with a diameter of about 1,400,000 kilometers (870,000 mi), it is surrounded by a very tenuous atmosphere that extends several solar diameters from the surface. It generates a tremendous amount of heat (about 5.6×10^{27} calories are radiated every minute, or 3.9×10^{23} kilowatts of power), but the earth intercepts less than one part in two billion of this total. Measurements made on the earth indicate that the rate at which energy impinges on a surface perpendicular to the sun's rays at the mean solar-earth distance is about 2.00 cal/cm^2/min. (1395 watt/m^2). This value is known as the solar constant, although no one is certain exactly how "constant" the output of the sun is. Variations of the total energy output are probably smaller than the accuracy of the measurements, which is about ± 2 percent. The intermittent outbursts of small particles and very short radiation, associated with disturbances on the sun, are largely absorbed in the outermost layers of the atmosphere.

According to Bethe's theory, the energy radiated from the sun is created through a complex thermonuclear reaction that converts protons (hydrogen nuclei) to alpha particles (helium nuclei). In the process, mass is converted to energy (in accordance with Einstein's familiar relationship, $E = mc^2$). The gravitational contraction of the enormous mass of the sun produces the temperature needed to make such a reaction possible. Although the sun is now converting mass to energy at the rate of about 4 × 10^6 tons per second, judging from the number of protons still available, the sun should continue shining for about 10^{11} years.

The radiant energy transmitted by the sun covers a broad range of the electromagnetic spectrum, from the very short gamma and X rays to radio wavelengths (Figure 3.2). The rate at which energy is emitted from each square centimeter of surface as a function of wavelength is very much like that for an ideal or black body at 6000 K, shown in Figure 3.3. Eighty percent of the solar energy falls in the visible (0.38 − 0.72 μm*) and near infrared (0.72 − 1.5 μm) portions of the spectrum, with the peak energy occurring at a wavelength of approximatey 0.56 μm, which is a blue-green. In contrast, a body of 300 K (27°C, which is about a dozen degrees warmer than the mean temperature of the air near the earth's surface) radiates energy at a rate of 1/160,000 that of a body at 6000 K, almost entirely in the far infrared, with a maximum emission near 10 μm wavelength.†

When the energy in the form of electromagnetic waves encounters the earth's atmosphere, some of the rays pass undisturbed, some are absorbed by the atmosphere, and the rest are turned back. We shall examine the ways in which all three of these occur.

(1) Absorption Oxygen, ozone, water vapor, carbon dioxide, and dust particles are the most significant absorbers of the "short-wave" radiation from the hot sun and the "long-wave" radiation of the cool earth. The gases are *selective* absorbers, meaning that they absorb strongly in some wavelengths, weakly in others, and hardly at all in still others. The very short ultraviolet (less than 0.20 μm) radiation of the sun is absorbed as it encounters and splits molecular oxygen into two atoms in the upper levels of the atmosphere. Ozone, formed by the combination of O and O_2, effectively absorbs ultraviolet light of longer wavelengths—those between 0.22 and 0.29 μm. The top portion of Figure 3.4 illustrates the relative effectiveness of oxygen and ozone as absorbers. Absorptivity is the fractional part of incident radiation that is absorbed. It can be seen that oxygen and ozone absorb almost 100 percent of all radiation at wavelengths less than 0.29 μm. For this reason, only a minute portion of the sun's ultraviolet radiation penetrates to the lower levels of the atmosphere. In the longer wavelengths neither of these gases absorbs very much energy, except for a narrow band (near 9.6 μm) in the infrared. About 2 percent of the sun's total radiation received on earth is depleted by ozone.

Water vapor is a significant absorber of radiation. Its complicated absorptivity characteristics are illustrated in Figure 3.4(b). Although not effective at wavelengths below 0.8 μm, where most of the *solar* radiation exists, it absorbs (strongly

---

*Wavelengths are usually given in micrometers, abbreviated μm, or in angstrom units, abbreviated Å. One micrometer is 10^{-4} cm, or 0.0001 cm, and 1 Å = 10^{-8} cm, or 0.00000001 cm.

†Planck's law, which gives the amount of energy radiated by a body as a function of temperature and wavelength, implies that the rate at which a body radiates energy increases with the fourth power of the absolute temperature (σT^4). The wavelength at which a body emits most intensely is given by Wien's law and is inversely proportional to the temperature. Thus, hot bodies not only radiate much more energy than cold bodies, but they do it at shorter wavelengths. "Red hot" is not as hot as "blue hot."

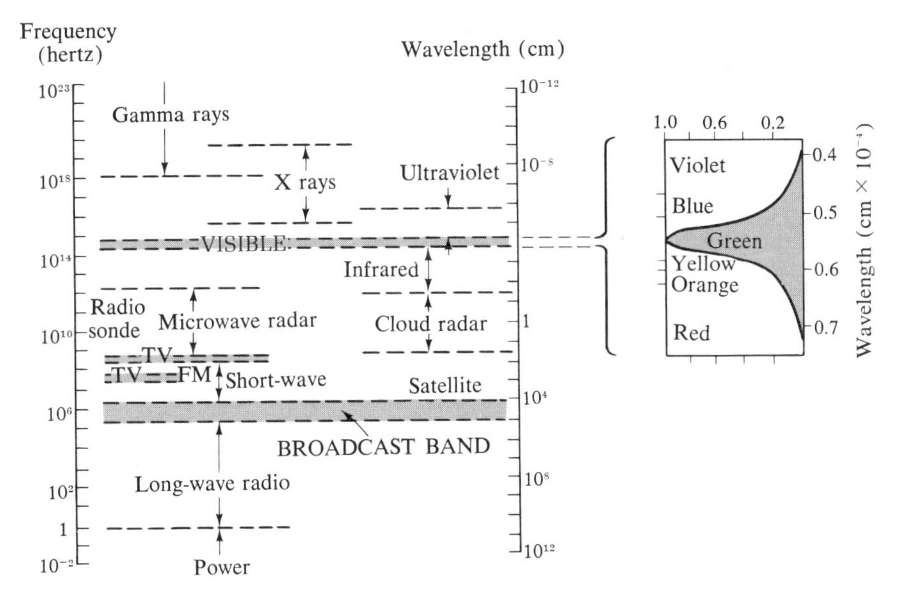

FIGURE 3.2 *The electromagnetic spectrum. The curve at the right gives the relative sensitivity of the human eye to electromagnetic radiation (maximum sensitivity at 0.555 × 10^{-4} cm wavelength).*

between 5 and 7 μm and moderately well beyond 15 μm) wavelengths at which the cool earth and its atmosphere emit much of their energy (Figure 3.3).

In summary, the atmosphere is essentially transparent between 0.3 and 0.8 μm, where most of the solar (short-wave) radiation occurs. But between 0.8 and 20 μm, where much of the terrestrial (long-wave) radiation is emitted, there are several bands of moderate absorptivity by water vapor.

(2) Scattering The atmosphere is composed of many, many discrete particles — gas molecules, dust, water droplets, etc. — but the empty space between particles is actually greater than the volume occupied by the particles. Each particle acts as an obstacle in the path of radiant energy (e.g., light waves) traveling through the atmosphere, much as rocks in a lake impede the progress of ripples in the water. The wave fronts are deformed by these obstacles into a pattern that makes it appear that the rays emanate from the obstacles. Thus, radiant energy propagating in a single direction is dispersed in all directions as it encounters each particle in its path. This dispersion of the energy is called *scattering.*

The effectiveness of a particle as a scattering center depends on its volume. For particles the size of gas molecules, the amount of scattering is much higher for the short wavelengths of light (blue) than for the long waves (red). It is for this reason that the white light of the sun and the moon becomes yellow or red on the horizon, while the sky, which is lit by scattered light, is blue. Astronauts have observed, as they ascend through the atmosphere, that the sky becomes darker, finally becoming black, as the density of the scattering particles decreases.

When the atmosphere contains many large dust particles or minute water droplets (haze), scattering is no longer very selective in terms of wavelength. The

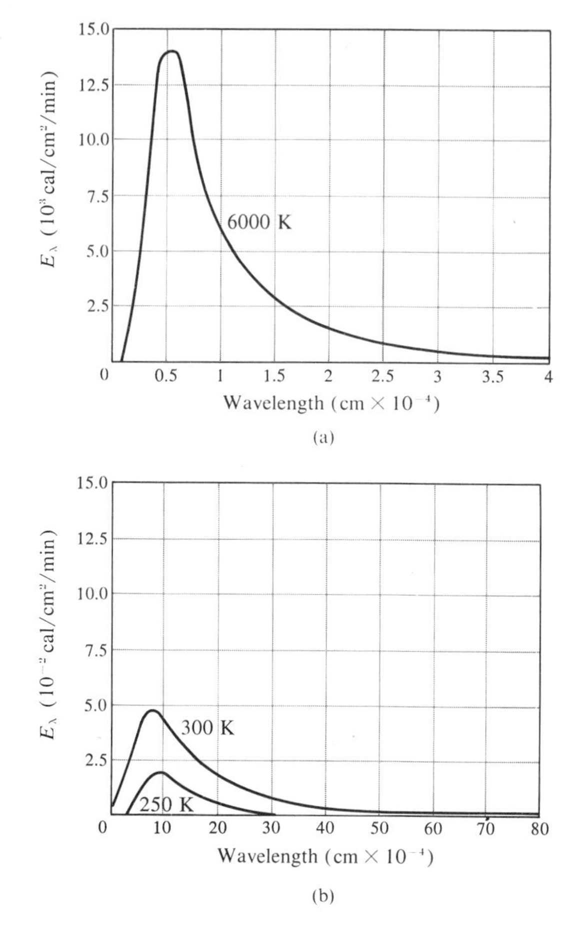

FIGURE 3.3 *Black-body emission of (a) a hot body such as the sun, and (b) a cool body such as the earth. [In comparing the curves, note that the vertical scale of (a) is 100,000 times that of (b).]*

long waves are scattered almost as much as the short waves, and the resulting sky color becomes less bluish and more white or milky. In fact, the blueness of a cloudless sky is an indication of its "purity;" i.e., how free it is of smoke, dust, and haze.

On the average, about 12 percent of the sun's radiation striking the earth's atmosphere is scattered; half of the scattered radiation is lost to space.

(3) Reflection The radiant energy from the sun encounters still another obstacle before it reaches the surface of the earth — clouds. Most clouds are very good reflectors but poor absorbers of radiant energy. The reflection by clouds depends primarily on their thickness, but also to some extent on the nature of the cloud particles (i.e., whether ice or water) and the size of these particles. The reflectivity of clouds varies from less than 25 percent to more than 80 percent, depending on the cloud thickness. Clouds absorb very little of the radiation that strikes them (not more than 10 percent), so that most of what is not reflected is transmitted through them.

In general, the earth's surface is a poor reflector of solar radiation, although the amount reflected varies greatly with the nature of the surface. Table 3.1 gives the reflectivity for a number of common surfaces.

TABLE 3.1 *Reflectivity or "Albedo" of Various Surfaces*

Surface	*Percent reflected*
Clouds (stratus) <500 ft thick	25–63
500–1000 ft thick	45–75
1000–2000 ft thick	59–84
Average of all types and thicknesses:	50–55
Concrete	17–27
Crops, green	5–25
Forest, green	5–10
Meadows, green	5–25
Ploughed field, moist	14–17
Road, black top	5–10
Sand, white	30–60
Snow, fresh fallen	80–90
Snow, old	45–70
Soil, dark	5–15
Soil, light, (or desert)	25–30
Water	8*

*Typical value for water surface, but the reflectivity increases sharply from less than 5 percent when the sun's altitude above the horizon is greater than 30°, to more than 60 percent when the altitude is less than 3°. Rough seas have a somewhat lower albedo than calm seas.

Solar energy absorbed by earth

Figure 3.5 presents a summary of what happens, on the average, to the solar radiation intercepted by the earth. Of the 100 percent of the total solar radiation reaching the top of the earth's atmosphere, approximately 20 percent is absorbed by the dust, water vapor, and cloud droplets in the atmosphere. About 5 percent is scattered by the atmosphere back to space and 3 percent is reflected to space from the earth's surface, which is a poor reflector. On the other hand, clouds are relatively good reflectors, and, on the average, they reflect 22 percent of the sun's radiation back to space. Because clouds play such an important role in the global energy budget, variations in the average cloud cover are potentially important in producing climate changes.

The earth's surface absorbs about 50 percent of the solar radiation: some of it (20 percent) comes directly from the sun, some (24 percent) after reflection by clouds, and the rest (6 percent) after being scattered by the air. The reflectivity of the earth's surface varies greatly, of course. Some fresh snow fields and water surfaces (when the sun is close to the horizon) reflect 90 percent or more of the incident rays. But a forest may reflect less than 10 percent and green grass fields only 10–15 percent.

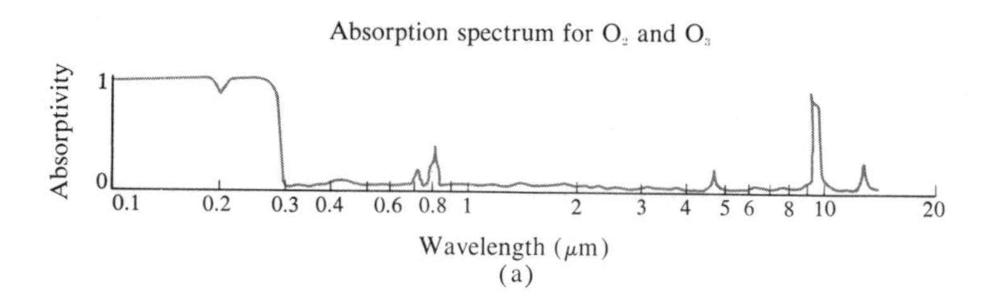

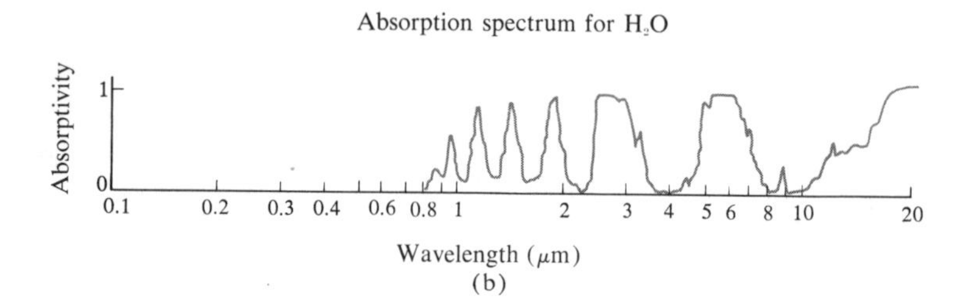

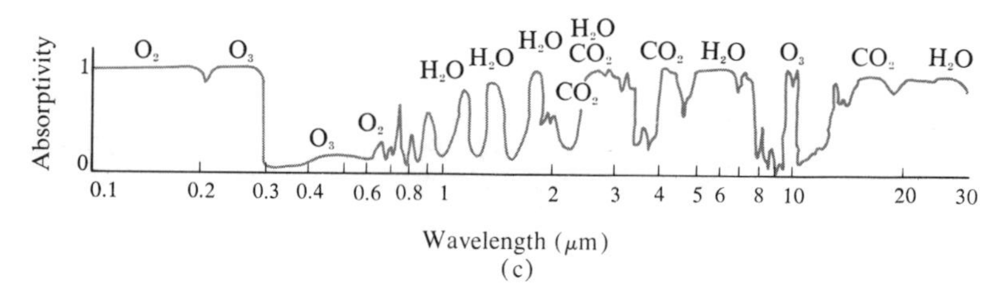

FIGURE 3.4 ***Absorption of radiation at various wavelengths by (a) O_2 and O_3; (b) H_2O; and (c) the principal absorbing gases.***

Thus, of the total energy arriving from the sun (2 cal/cm²/min), approximately 70 percent is absorbed by the earth's surface and atmosphere. The rest, 30 percent, is lost to space, having been reflected by clouds and the earth's surface or scattered by the particles in the air. The average reflectivity, "whiteness," or *albedo* of the earth is said, therefore, to be 0.30, since that is the fractional part of the incident radiation that is bounced off the earth. In contrast, the moon's albedo is only about 7 percent, which means that it is not nearly as bright as the earth.

3.3 The Earth's Heat Balance

Studies indicate that over moderately long periods of time (between hundreds and thousands of years) the mean temperature of the earth is essentially constant. This indicates that there exists a long-term balance between the earth and space. It follows, then, that since 70 percent of the solar energy striking the earth is absorbed, an equal amount must be reradiated to space. Figure 3.5 shows what happens to the

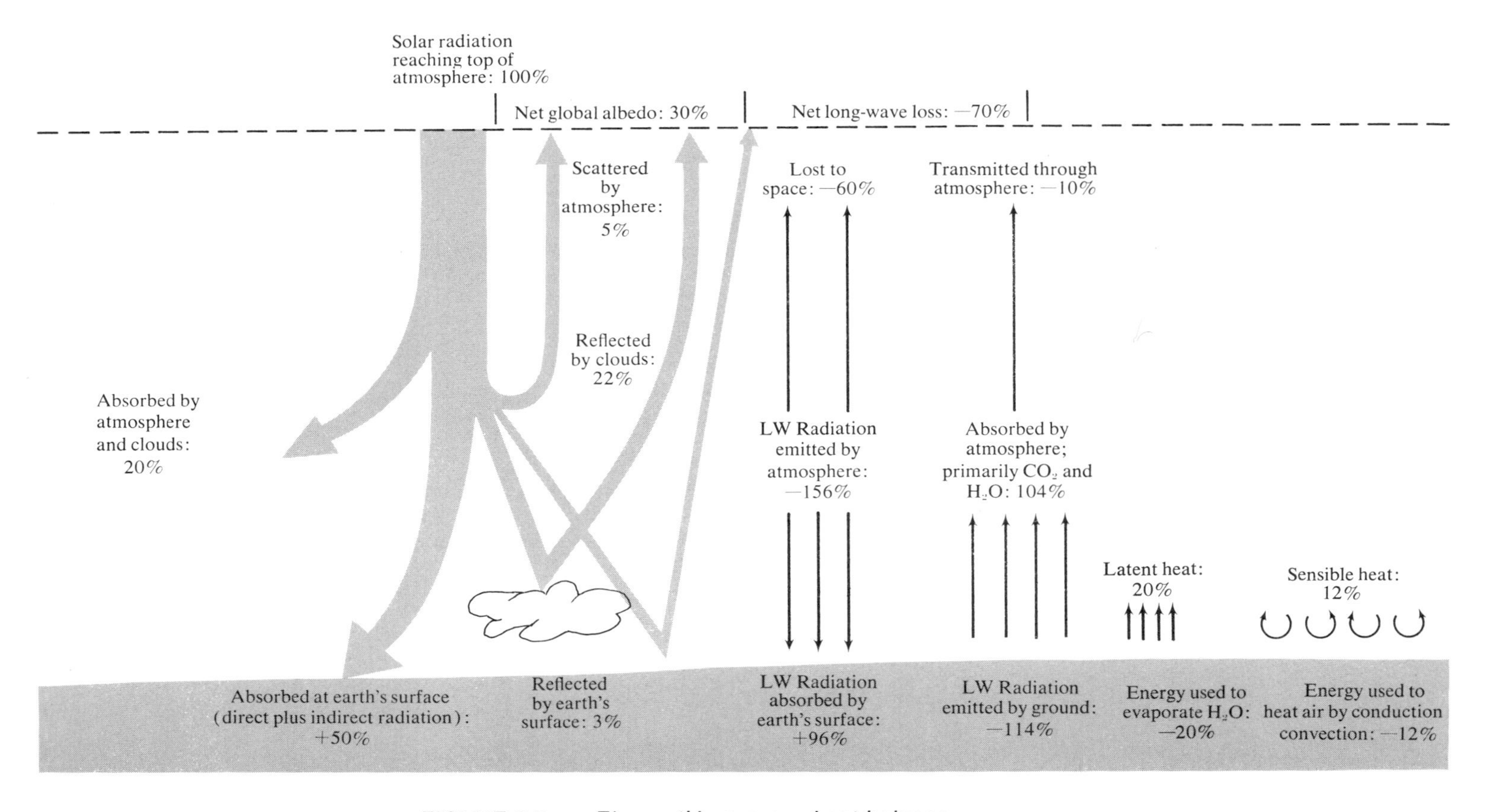

FIGURE 3.5 *The earth's average heat balance.*

earth's energy. Note that 70 percent is lost by the earth and its atmosphere to space. Keep in mind, though, that although there is a heat balance for the planet as a whole, all parts of the earth and its atmosphere are not in radiative balance. In fact, it is the imbalance between incoming and outgoing energy over the earth that leads to the creation of wind systems that act to alleviate the surpluses and deficits of heat that would otherwise result.

Of special interest in the long-wave energy transfers is the fact that the amounts emitted by the earth and the air actually exceed the total solar energy amount retained by the earth (70 percent). This can be explained in terms of the "blanketing" effect of the atmosphere, which keeps the earth's surface and lower layers of the atmosphere a good deal warmer than they would be without the atmosphere. For example, the moon's sunny surface, which absorbs almost twice as much energy per unit area as does the earth's surface, is more than 20°C colder because it lacks an atmospheric "blanket."

Two gases—water vapor and carbon dioxide—play the most important role in keeping the earth warm. Except for the "windows" between about 8 and 13 μm (Figure 3.4), these gases block the direct escape of the infrared energy emitted by the earth's surface. Although the atmosphere absorbs only a small percentage of the short-wave solar radiation (20 percent, Figure 3.5), it is quite opaque to the long-wave terrestrial radiation (absorbing $104/114 = 91$ percent, Figure 3.5). Only when the earth's surface temperature is fairly high does the radiational loss through the transparent bands and from the top of the atmosphere equal the amount absorbed from the sun. This heat-retaining behavior of the atmosphere is somewhat analogous to what happens in a greenhouse. In a greenhouse, the glass (or more recently, plastic) roof and sides permit the sun's energy to enter and be absorbed by plants and earth, but they prevent much of the interior heat from escaping by blocking mixing of the inside air with that on the outside. To a lesser extent, the glass also prevents escape of the energy radiated at long wavelengths from inside the greenhouse. The role of moisture in the atmosphere, allowing solar energy to pass through while absorbing much of the earth's long-wave radiation, has thus come to be known as the *greenhouse effect.* This effect is quite noticeable when one compares the rapid temperature fall at night in the desert, where the air is dry, with the slower decrease in temperature in coastal regions, where the air is moist. On a global basis, the effectiveness of the atmospheric "greenhouse" is illustrated by the fact that the mean air temperature near the surface is about 13°C, while the overall "planetary temperature" is only about −20°C. Venus's greenhouse effect is even more dramatic. Its radiative temperature is about −30°C, but its surface temperature is estimated to be more than 450°C.

3.4 Distribution of the Earth's Heat Energy

As mentioned above, although the overall income and outgo of radiant energy are essentially in balance, they are not in balance everywhere on the earth. This is mainly because the amount of incoming energy varies greatly from place to place. It is also caused, to a lesser extent, by variations in the intensity of outgoing radiation.

The amount of energy *emitted* by the earth and the atmosphere to space is controlled largely by the amount of moisture in the air; the distribution of solar energy *absorbed* by the earth's surface is controlled mostly by the earth's movements, the distribution of physical properties of the surface, and cloudiness.

The latitudinal variations of absorbed solar energy can be easily understood if one bears in mind three facts: (1) the earth is essentially a sphere, (2) the sun is so far away that its rays of light are approximately parallel, and (3) the earth is rotating. Only one-half of the sphere can be illuminated at one time, and the angle that the sun's rays make with the sphere's surface will decrease from 90 degrees at the exact center of the lit hemisphere to 0 degrees at the edges (where the shadow begins). This is illustrated in Figure 3.6(a). Angle *c* equals 90 degrees at the one point where the sun's rays are perpendicular to the surface, while angle *b* is less than 90 degrees, and angle *a* is less than angle *b*. The rays entering with the angle *a* will, of course, traverse a greater mass of atmosphere than those entering with angle *b* or angle *c* (since distance $d_1 > d_2 > d_3$), and therefore they will be subject to greater depletion by absorption, reflection, and scattering. But even more important, the intensity will be less at lower solar angles because the same amount of energy will intercept larger surface areas, as is illustrated in Figure 3.6(b), so that the amount of energy received by each square kilometer of surface will be less for low angles of the sun than for high ones.

Finally, the earth rotates about an axis, and the half of the sphere being illuminated is continuously changing. Consider the ring formed by the circular edge of the earth's shadow. When the axis of rotation lies in the same plane as this ring, it is evident that the ring will divide each latitude circle exactly in half. Thus, all places will have precisely 12 hours of sunlight and 12 hours of darkness each day. But when the axis of rotation does not fall in the plane of the shadow's ring, the latitude circles will not be divided into two equal parts, except at the equator. The ratio of night to day at each latitude will be in the same proportion as the two segments of the latitude circle created by the shadow ring.

As the earth moves in its annual course around the sun, the axis of rotation sometimes coincides with the shadow ring—and sometimes it does not. The axis of rotation is not perpendicular to the plane containing the path of the earth's revolution, but rather it is tilted at an angle of 23½ degrees from the normal. This means that only at the equinoxes (March 21 and Sept. 23, approximately) does the axis lie in the plane of the shadow ring.

Figure 3.7 shows the way in which the total energy (*insolation*) received each day varies with time of year and latitude. Note that in the summer, in both hemispheres, the total daily energy received varies little between the poles and the equator; this is because the lower solar angles in the polar regions are compensated for by the greater duration of sunshine each day. In the winter, of course, the latitudinal variation in the amount of energy received is very great, since practically none is received at high latitudes while the equatorial region's supply remains almost unchanged throughout the year.

There is a slight difference between the two hemispheres in the distribution of energy received throughout the year. You will note from Figure 3.7 that the Southern Hemisphere receives a little more energy in its summer than does the Northern

Hemisphere in its summer, and conversely during their respective winters. The small differences occur because the earth moves in a slightly elliptical path (Figure 3.8) about the sun, and at the beginning of January the sun and the earth are

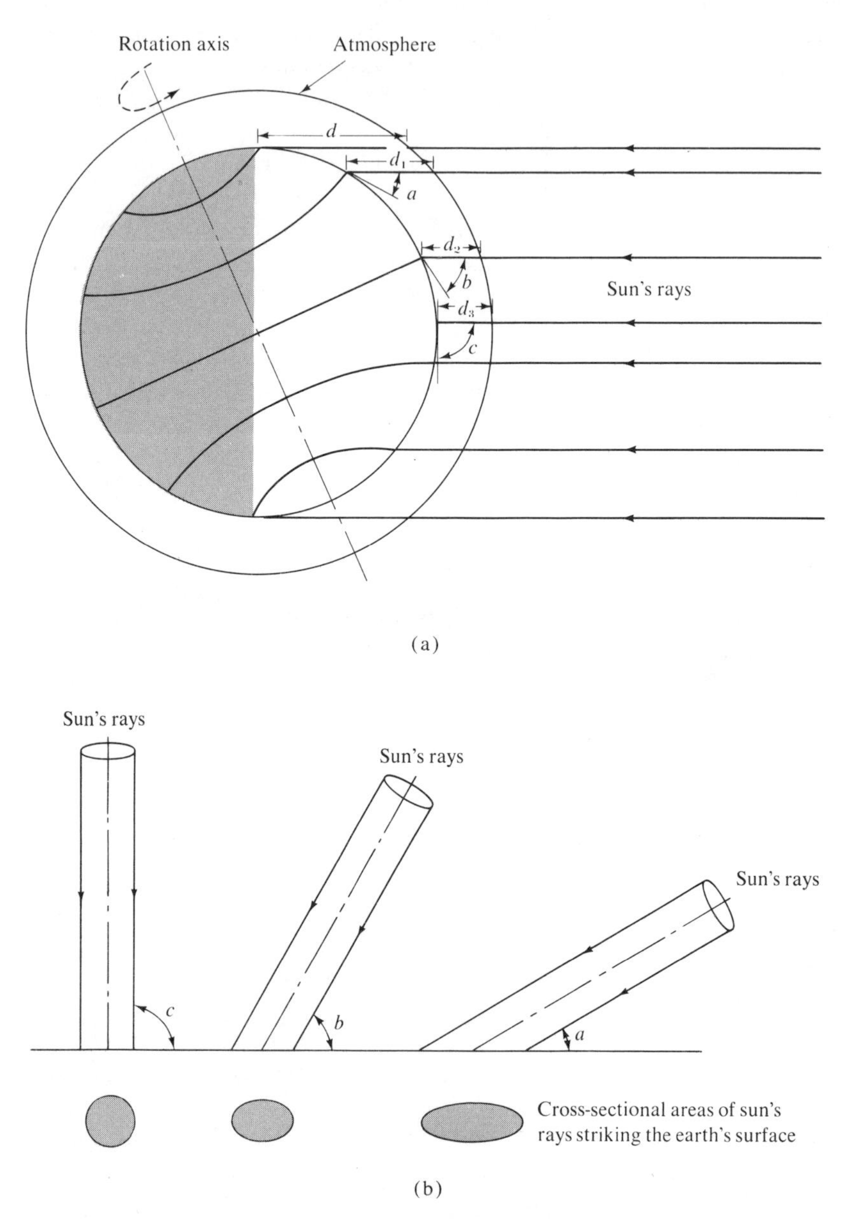

FIGURE 3.6 *The intensity of solar radiation depends on the angle at which the sun's rays strike the earth's surface. (a) The angles of incidence a, b, c; and the depths of penetration through the atmosphere, d_1, d_2, d_3, at different parallels of latitude. (b) The varying cross-sectional area on the earth's surface due to different angles of incidence.*

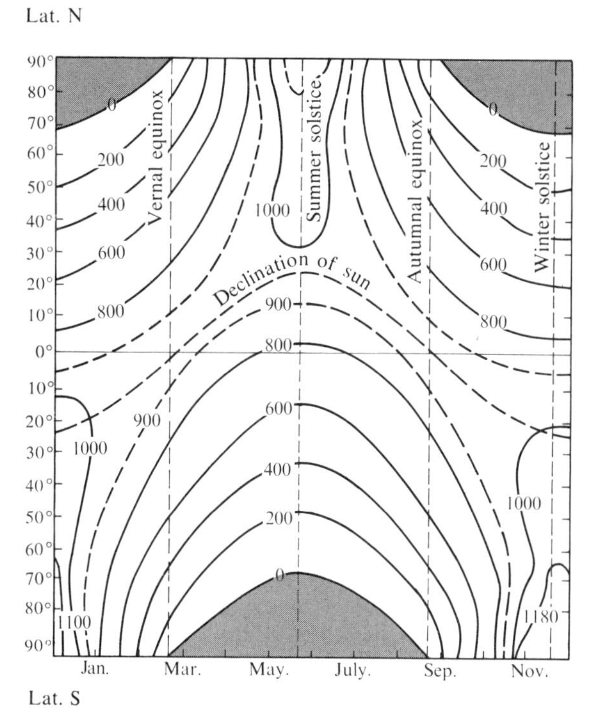

FIGURE 3.7 *Undepleted insolation in cal/cm²/day as a function of latitude and date. Shaded areas represent latitudes entirely within the earth's shadow.*

closest together (*perihelion*). The difference in total energy received by the earth between *aphelion* and perihelion is only 7 percent.

The amount of incoming and outgoing energy, averaged over the entire year, is shown for each latitude in Figure 3.9. It can be seen that, as would be expected, the incoming short-wave energy decreases a great deal between the equator and the poles, while the outgoing long-wave energy is nearly constant. Averaged over all latitudes, the incoming energy (curve I) equals the outgoing energy (curve II). But there is

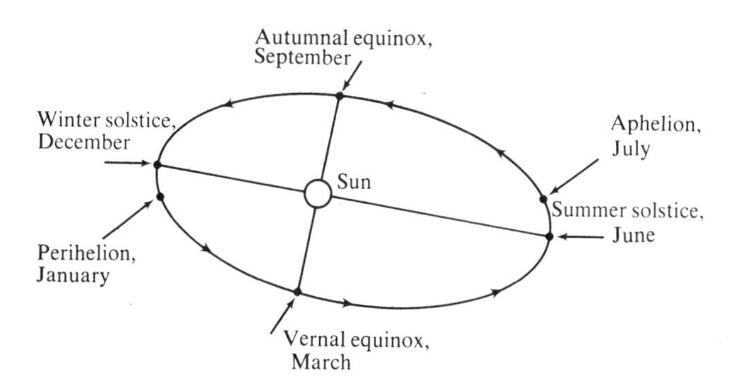

FIGURE 3.8 *Elliptical orbit of the earth (eccentricity exaggerated).*

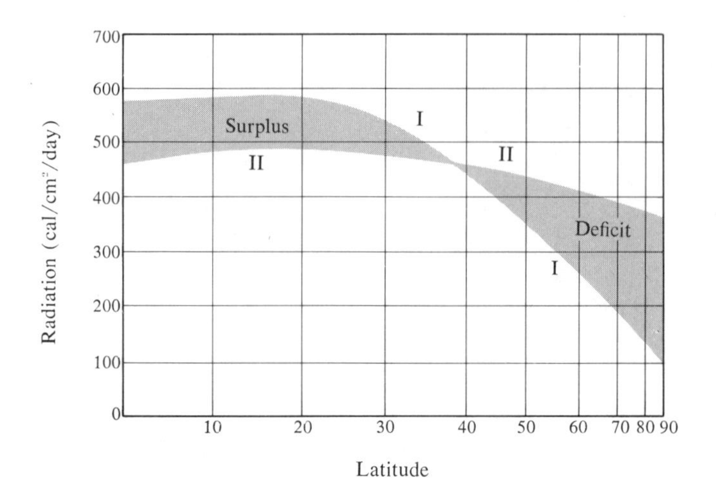

FIGURE 3.9 *Curves I and II represent mean annual insolation and outgoing long-wave flux, respectively, at the tropopause.*

a surplus of energy income at tropical latitudes and a deficit in the polar regions. If air movements (and, to a lesser extent, ocean currents) did not exist to redistribute the energy, the poles would become steadily colder and the tropics steadily warmer, until a new radiative equilibrium was established.

If there were only latitudinal variations in the energy received and absorbed by the earth and its atmosphere, meteorology would be a considerably simpler study. However, the absorptive properties of the air and the surface of the earth are not distributed in a smooth, unchanging pattern. The radiation absorption properties of the atmosphere depend on the clouds, moisture, and dust in the air, the concentrations of which may vary both geographically and with time. Surface properties are also erratically distributed over the face of the earth, and even they change with time.

The most pronounced differences in surface thermal properties are those between land and sea. Under identical insolation conditions (same solar angles, duration of daylight, atmospheric transparency), the temperature changes experienced by the water will be much less than those of the land surface. The principal reason for this is that water is a fluid and so can be mixed; as a result, its heat tends to be distributed over a much greater mass than is the case with "stagnant" land. The heat absorbed by a land surface tends to be confined in the upper few inches, while in water the heat may be distributed to depths of hundreds of feet. There are other reasons for the smaller temperature range of water surfaces: (1) because water is transparent, radiation can penerate to depths of tens or even hundreds of meters, so that the energy is absorbed by a great mass of water; (2) water generally has a higher specific heat than does land (i.e., more heat is required to raise the temperature of a gram of water 1°C than for land); and (3) some heat is used in evaporation of water (latent heat).

The fact that the oceans act as heat reservoirs is illustrated by the January and July mean air temperature maps of Figure 3.10. Note how much more the tempera-

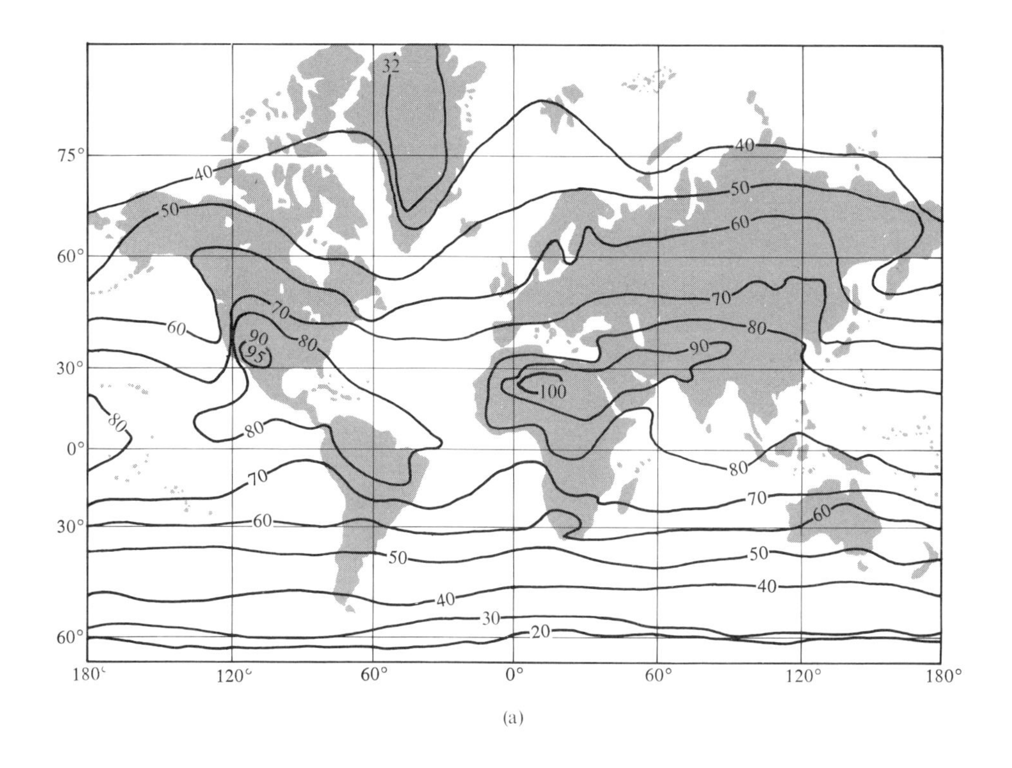

(a)

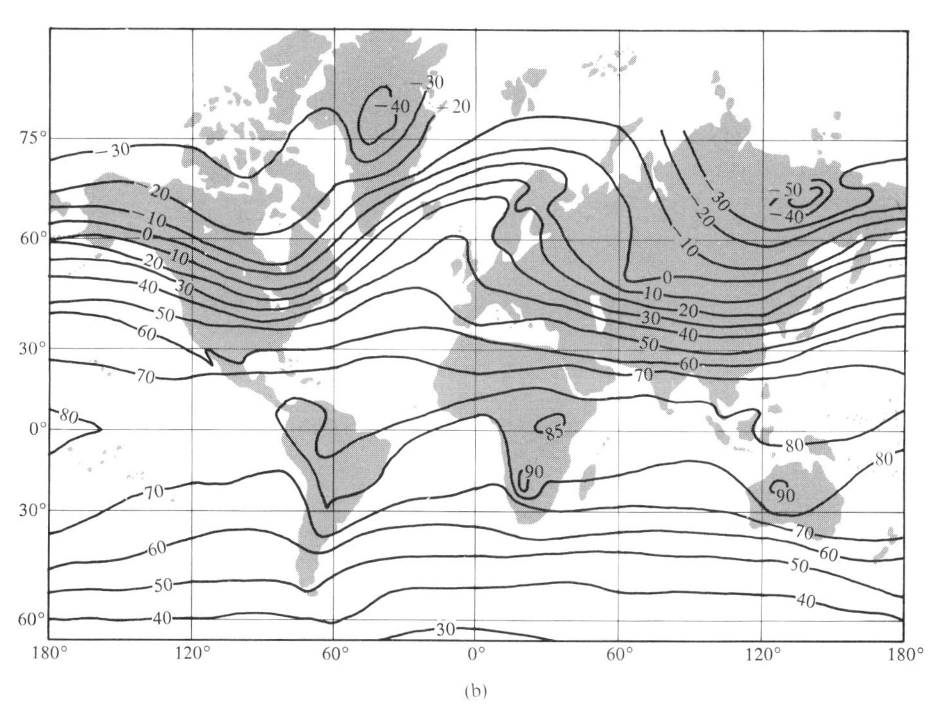

(b)

FIGURE 3.10 *(a) World distribution of mean temperature (°F) for July; (b) World distribution of mean temperature (°F) for January.*

ture varies between seasons in middle and high latitudes over the continents than it does over the oceans. For example, at latitude 45°N, the annual range of temperature over the continents is about 60°F, while over the Pacific Ocean at the same latitude, the range is only about 10°F. Note also how the isotherms dip equatorward over the oceans in summer and poleward in the winter, indicating that the ocean is cooler than the land in summer and warmer than the land in winter.

Temperature lag

The times of high and low air temperatures do not coincide with the times of maximum and minimum solar radiation, either on an annual or daily basis. The months of July and August are generally the hottest of the year, while January and February the coldest; yet the greatest intensity of radiation occurs in June and the lowest in December. On a daily basis, the highest temperature normally occurs at 2 or 3 P.M., yet the greatest intensity of insolation each day occurs near noon.

This lag in temperature is explained by Figure 3.11. The earth loses heat continuously through radiation. During some months of the year and some hours of

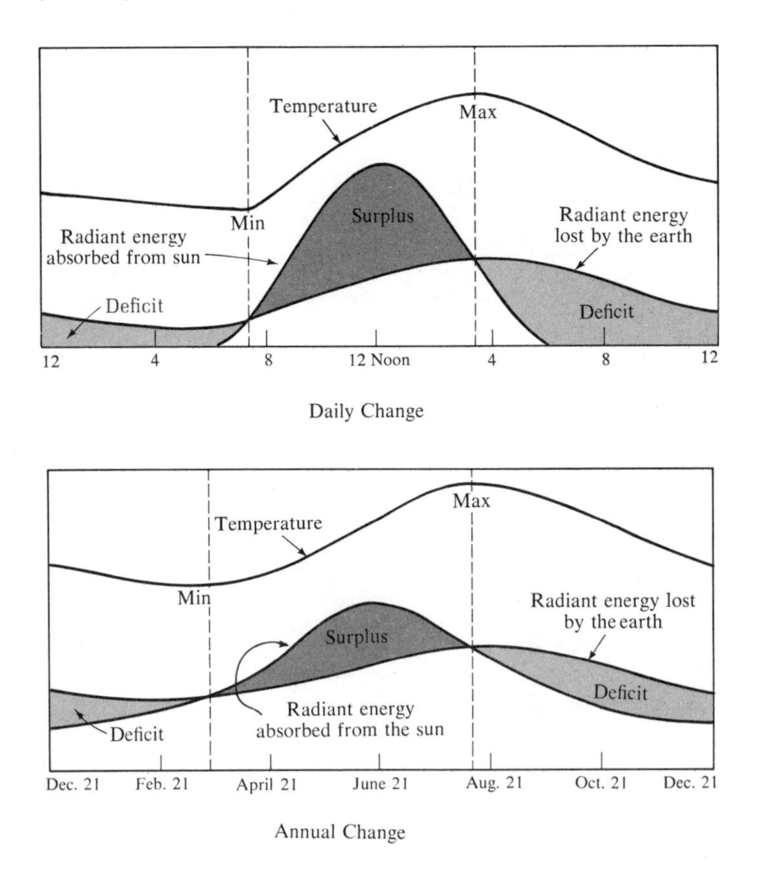

FIGURE 3.11 *The daily and annual temperature maxima and minima lag behind the maxima and minima of solar radiation.*

the day, the incoming energy exceeds the outgoing energy of the earth. While this is occurring, the temperature will be increasing, since the air's heat content will be rising. The maximum temperature will occur at the time when the incoming energy ceases to exceed the outgoing. Thereafter, when the outgoing energy is greater than the incoming, the temperature will fall until the two are again in balance. At the point where a "surplus" of energy begins to appear, the lowest temperature will have occurred.

3.5 The Possible Use of Solar Energy

Recent concern over the increased cost and limited supply of oil and other fossil fuels as an energy source has revived interest in the possibilities of the direct use of solar energy. Both the potential and some of the limiting factors are evident from the discussion in the previous sections of this chapter: An average of about 0.5 $cal/cm^2/min$ is intercepted by the entire earth (see Problem 1 at end of chapter); of this amount, 50 percent (Figure 3.5), or 0.25 $cal/cm^2/min$, reaches the surface of the earth. This comes to 1.74 watts on every square meter of the earth's surface (see Appendix 1 for conversion of units), or more than 4 kilowatt-hours per day. This amount of energy is about half as much as is required to heat and cool the average U.S. house. In other terms, if the 4 kilowatt-hours per square meter could be converted to electricity with an efficiency of 10 percent, some 400,000 kilowatt-hours of electricity could be obtained from a solar collection area of 1 square kilometer; this represents about 1/75 (1.33 percent) of the energy consumed by a city with a population of from 1 to 1½ million. All of the U.S. electrical energy could be supplied with solar energy collectors covering 0.14 percent of the land, again assuming 10 percent efficiency.

Although these computations are correct, they are based on the *average* incidence of solar energy on the earth's surface. The amount of sunshine intercepted varies greatly with latitude and season (Figure 3.7) and time of day; the amount reaching the surface depends, of course, on the atmospheric transparency—primarily cloudiness—which varies greatly from day-to-day. Thus, one of the greatest problems in using solar energy is its intermittency. Complete reliance on solar energy requires the development of techniques for storing energy. This will probably involve the creation of synthetic fuels from solar energy.

The heating of buildings and water with solar energy already exists and is relatively simple. A dark radiation-absorbing surface is exposed to the sun's rays, perhaps mounted on sloping roofs; panes of glass are mounted above the surface to create a greenhouse effect, blocking some of infrared radiation loss from the heated surface. Air or water pumped over the heated surface transport the heat to the interior of the building or to water tanks.

There are many other "indirect" solar energy sources that are often suggested for "tapping," such as the conversion of plant material to fuels, the harnessing of ocean tides and waves, and of winds. Each of these potential sources contains only a very small proportion of the total solar power reaching the earth's surface. A sample calculation of the energy extractable from the wind is presented in Problem 11, at the end of this chapter. The greatest obstacle to the exploitation of both direct and indirect solar energy is the cost of development.

PROBLEMS

1. Compute the total energy per minute intercepted by the earth and the fraction of the total solar energy output this represents. (Hint: The cross-sectional area of the earth = $\pi r^2 = 3.1416 \times (6000)^2$ km^2, while the surface area of an imaginary sphere surrounding the sun at the distance of the earth from the sun = $4\pi d^2 = 4 \times 3.1416 \times (149.6 \times 10^6)^2$ km^2. (The solar constant = 1395 W/m^2.)
2. If you were attempting to observe the temperature distribution on the moon's surface by measuring the infrared radiation, what wavelength band would give the best results, considering the atmosphere's transparency?
3. Compute the elevation angle of the sun at noon at the latitude of your city on Dec. 21. What will be the length of the daylight (not counting twilight)?
4. Explain why nighttime temperatures are generally lower on nights when the humidity is low than when it is high.
5. What is the essential difference between the phrases "light from the sun" and "radiation from the sun"?
6. What angle between the earth's axis of rotation and the plane containing the earth's path around the sun would provide (a) the least difference between seasons? (b) the greatest difference?
7. Suppose you wish to design your living room so that it receives as much sunshine as possible during the winter, and as little as possible during the summer. If you live in the Northern Hemisphere, in which direction should the windows face? What if you live in the Southern Hemisphere?
8. How do you explain the difference in albedo between the earth and the moon? Is the dark side of the moon illuminated by "earthlight," as we are by moonlight?
9. How much heat per unit area is required to melt a layer of snow at a rate of 1 inch/day, assuming a liquid water equivalent of 10:1? What percentage of the mean total outgoing radiation at latitude 50 degrees does this represent?
10. The efficiency of an engine is defined as the ratio of the rate at which work is extracted to the rate at which energy is added. Use the results of Problem 1 and the fact from Figure 3.5 that 70 percent of the incident solar radiation is absorbed by the earth and its atmosphere to show that the average rate of energy addition to the earth–atmosphere system is 244 W/m^2. If the rate at which kinetic energy is generated in the atmosphere is about 2 W/m^2, how efficient is the atmospheric heat engine? How does this compare with the average automobile engine?
11. Wind power has long been used to do a very minor portion of people's work. This is not too difficult to understand. Suppose we have a windmill that always faces into the wind and has a radius of 5 meters (area = 78.5 m^2). It drives an electric generator. Assuming an average wind speed of 5 m/s (11.2 mi/h), the air's energy per gram is $\frac{1}{2}V^2 = 12.5$ m^2/s^2. The volume of air intercepted by our windmill in one year (3.1536×10^7 s) equals the distance traveled by the air times the windmill area = (5 m/s) $\times$ (3.1536×10^7 s) $\times$ (78.5 m^2) = 1.2378×10^{10} m^3. Since the average

air density is 1.2 kg/m^3, the number of kilograms striking the blades is $(1.2378 \times 10^{10}$ m$^3) \times (1.2$ kg/m$^3) = 1.485 \times 10^{10}$ kg. Thus the total possible energy is $(1.485 \times 10^{10}$ kg$) \times 12.5$ m^2/s$^2 = 18.56 \times 10^{10}$ kg m^2/s^2 or 18.56×10^{10} Joules. Since 1 Joule equals 2.777×10^{-7} kWh, the total possible energy is 5155.5 kWh. If the efficiency of energy capture and conversion to electricity were as high as 10 percent, the useful energy generated in one year would be 516 kWh. How does this compare with the average household consumption? Is it practical? What other factors (for example, characteristics of the wind) have not been considered?

12. The axis labeled "latitude" in Figure 3.9 is scaled as the cosine of latitude rather than linearly with latitude. Why?

Air in Motion

The uneven distribution of heat resulting from latitudinal variations in insolation and from differences in absorptivity of the earth's surface leads to air motions. The mechanics of this conversion of heat energy to kinetic energy will be the subject of this chapter.

We will be examining the deviations of air motion from those that are due to the planetary motion. The atmosphere as a whole follows the earth in its movements through space; it also rotates with the earth from west to east, so that at the equator the air moves eastward at a speed of more than 1600 km/h, while at latitude 60 degrees it moves eastward at half that speed. Of course, at the poles its eastward speed is zero. Because the ground moves at the same eastward speed, these motions go unnoticed by the earthbound observer. Winds are those motions of the air *relative* to the earth.

4.1 Principal Forces in the Atmosphere

According to *Newton's first law,* for a body to change its state of motion, it must be acted upon by an unbalanced force. There are two classes of forces that affect the atmosphere: (1) those that exist regardless of the state of motion of the air and (2) those that arise only *after* there is motion. The first category can be thought of as the fundamental or basic forces, since without them there would be no motion. These basic "driving" forces are produced by gravitational attraction and pressure. In the second group of forces are friction or "drag" and Coriolis forces.

The "driving" forces

The gravitational pull of the earth is always directed downward; the strength with which it acts on any "parcel" of fluid is proportional to the mass of the parcel. The only way a parcel of fluid can experience a net force due to pressure in some particular direction is if one "side" of the parcel is being acted on by a pressure different from that acting on the opposite "side." In Figure 4.1, the fluid in the pipe will remain stationary if the pressure, p_1 ($= F_1/A$), at the face of the piston on the right side is equal to the pressure, p_2 ($= F_2/A$), at the piston face on the left. Only if the two forces, F_1 and F_2 (or the two pressures, p_1 and p_2), *differ* will the

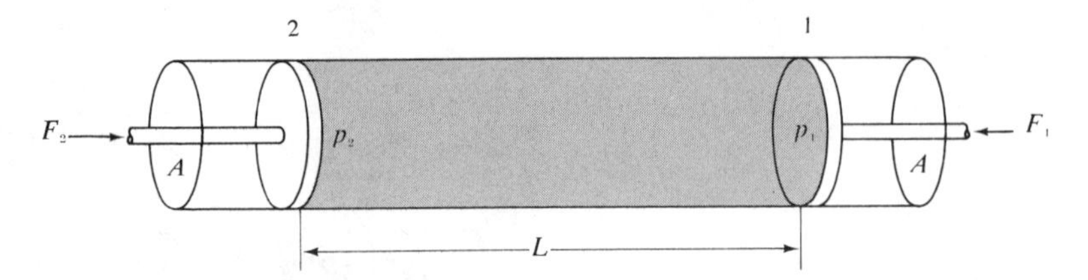

FIGURE 4.1 *Pressure gradient in a pipe. (A = cross-sectional area of pipe; L = distance between pistons.)*

fluid accelerate. According to *Newton's second law,* the magnitude of the *acceleration* experienced will then depend on the net force ($F_1 - F_2$) and on the mass being acted upon.

Assuming that the pipe has a cross-sectional area A, the volume of the mass of fluid on which the net force $F_1 - F_2$ is acting is given by $A \times L$ where L is the distance between the pistons. Therefore, the net force acting on each unit volume of fluid is

$$\frac{F_1 - F_2}{A \times L} = \frac{p_1 - p_2}{L} \tag{4.1}$$

Thus, the force per unit volume in a fluid is directly proportional to the pressure difference and inversely proportional to the distance. This ratio of pressure increment to distance, $(p_1 - p_2)/L$, is called the *pressure gradient,* since it measures the grade, or slope, of the pressure. The direction of the acceleration will be "down slope"; i.e., from high to low pressure. In Figure 4.1, if $p_1 > p_2$, the acceleration will be from right to left.

Of the two principal forces producing accelerations in the atmosphere—gravity and pressure gradient—the first *always* acts vertically downward, but the second can theoretically act in any direction. However, in the atmosphere the pressure gradient is almost entirely directed in the vertical also. Figure 4.2 illustrates a typical pressure pattern in a vertical cut of the atmosphere. In the figure the vertical component of the pressure gradient force, obtained by dividing the difference in pressure between points C and A by the vertical distance between these points, is 24 mb/200 m, or 120 mb/km. The horizontal component, obtained by dividing the difference in pressure between points E and D by the horizontal distance between these points, is 4 mb/400 km, or 0.01 mb/km. Thus the vertical component of the pressure gradient force is 12,000 times greater than the horizontal component in this typical example. Because the large vertical pressure gradient force, which is directed upward, is almost entirely balanced by gravity, which acts downward, it is customary to treat horizontal forces (and motions) separately from those in the vertical.

In the vertical, the pressure always decreases with height (since the mass of fluid remaining above must decrease with height), so the vertical acceleration of a parcel produced by the *pressure gradient* must *always* be directed upward. The gravitational pull on a parcel is, of course, *always* downward. It is the *net* difference between these oppositely directed forces that determines whether a parcel will accelerate upward or downward and at what rate.

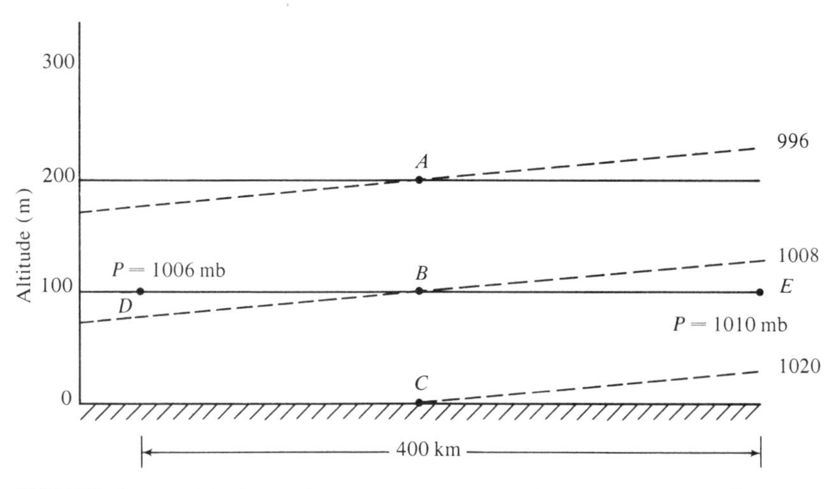

FIGURE 4.2 *Horizontal and vertical variation of pressure. Dashed lines are isobaric surfaces labeled in millibars (mb).*

The balance between the vertical component of the pressure gradient force and gravity is called *hydrostatic* balance. In Figure 4.3 we depict a parcel of air with surface area of the top and bottom faces A, depth Δz, and mass M. If we equate the net upward force on the parcel, $(P_1 - P_2) \times A$ with the downward gravitational force $M \times g$, we obtain

$$(P_1 - P_2) \times A = M \times g \tag{4.2}$$

Since the mass M equals the density ρ times the volume $A \times \Delta z$, (4.2) can be written as

$$P_1 - P_2 = g \times \rho \times \Delta z \tag{4.3}$$

which is the *hydrostatic equation.*

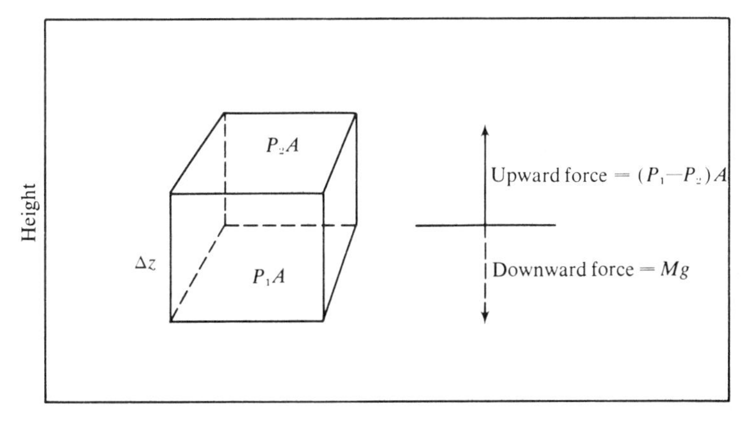

FIGURE 4.3 *Hydrostatic balance represents the balance between the vertical component of the pressure gradient force and gravity.*

The hydrostatic equation states the variation of pressure in the vertical is greater in cold air (high density) than in warm air (low density). In Figure 4.4, for example, the isobaric surfaces are closer together to the north where the density is high. Therefore, the high pressure (on a constant height surface) that exists near the surface in the north becomes a region of low pressure aloft. In contrast, the low near the surface in the south becomes a high aloft. This principle can be seen in the climatological distribution of pressure which shows high pressure at the surface over polar regions with low pressure aloft, while at the equator low pressure at the surface is overlain by high pressure aloft. On a smaller scale, cold anticyclones (highs) at the surface weaken with height and become cyclones (lows) aloft; while warm cyclones (such as hurricanes) at the surface weaken and become anticyclones aloft.

Thermal circulation

Consider the north-south cross sections of the atmosphere illustrated in Figure 4.4. If the air to the south is warmer than that in the north, the density at any level will increase from south to north. The less dense, warm air in the south will rise while the denser, cold air in the north sinks.

Near the surface, the pressure surfaces slope upward toward the colder, denser air of the north. However, the north-south slope of the isobars will decrease with altitude above the surface until eventually the slope will be reversed. This is because the vertical pressure gradient in *cold* air is greater than it is in *warm* air as shown by the hydrostatic equation (4.3). At any particular pressure, the density depends on the temperature [Equation of state (1.2)], warm air is less dense than cold air.*

Therefore, although the horizontal flow is always from high to low pressure, near the surface this means north to south (cold to warm) motion, while aloft it will be just the opposite. The closed circuit formed by the moving parcels of air is a *thermal circulation.* Note that it looks very much like the pattern of motion that occurs in a pan of water that is being heated at one point.

In summary, there are two rather large "basic" forces acting on the atmosphere—pressure gradient and gravity. The latter is directed entirely in the vertical, while the former has a very small component directed in the horizontal. Even though each of the two forces acting in the *vertical* is much greater than that in the *horizontal,* this does not mean that vertical motion is much stronger than horizontal motion; remember that it is *net* or unbalanced force that determines acceleration, and the two vertical forces are almost always very nearly equal and oppositely directed. Actually, except in small circulation cells such as those of a thunderstorm, the vertical air velocity is normally only a tenth or a hundredth of the horizontal velocity.

---

*The density also depends slightly on the moisture in the air. A mixture of air and water vapor is less dense than dry air because water has a lower molecular weight—18—than the average for dry air—29.

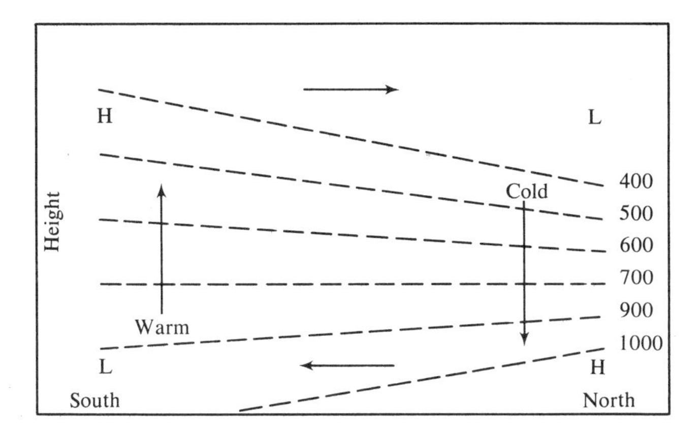

FIGURE 4.4 *Variations in the separation of isobaric surfaces between warm and cold air lead to reversal of pressure systems with height. Dashed lines are isobars (mb). Direction of circulation is indicated by arrows.*

4.2 Forces That Arise after There Is Motion

The effect of the earth's rotation

Large-scale flow in the earth's atmosphere does not follow the simple pattern of the thermal circulation shown in Figure 4.4. The horizontal wind blows more nearly *perpendicular* to the pressure gradient than along it. Figure 4.5 shows surface weather maps for the Northern and Southern Hemispheres, with arrows designating

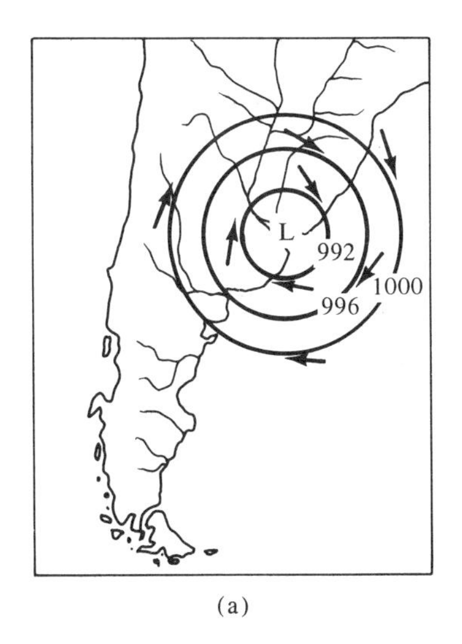

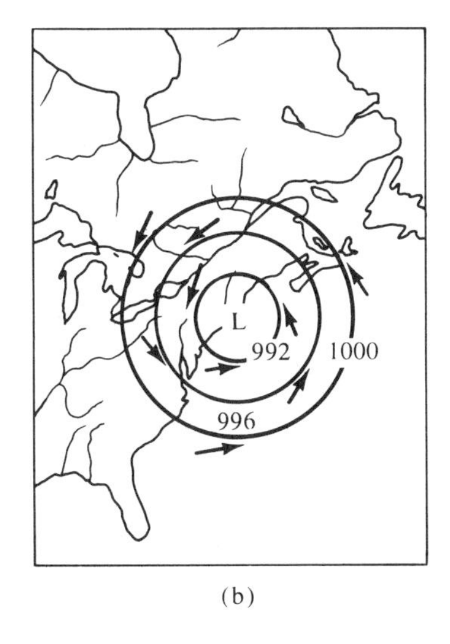

FIGURE 4.5 *Pattern of wind around an area of low pressure near the surface. (a) The Southern Hemisphere; (b) The Northern Hemisphere.*

the observed wind directions; note that the air is moving more or less *along* the isobars, rather than across them. Evidently something steers the air flow to the right of its target (low pressure) in the Northern Hemisphere and to the left in the Southern Hemisphere. The only thing that could reverse its effect between the two hemispheres is the earth's rotation. This is illustrated by the turntable of Figure 4.6. The top of the turntable has been given the same sense of rotation (counterclockwise) as the earth's Northern Hemisphere; however, if one looks at the same turntable from below, the sense of rotation is opposite (clockwise).

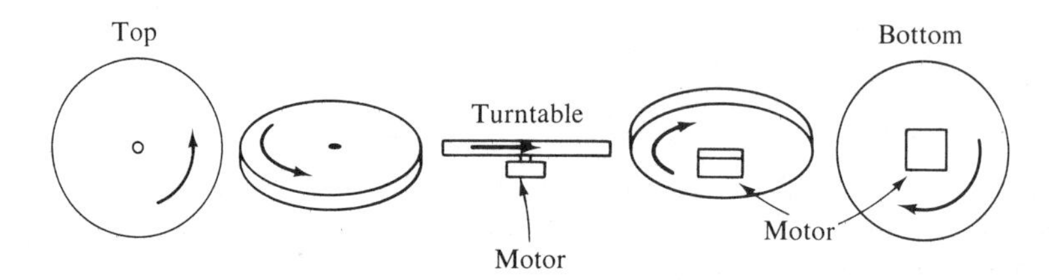

FIGURE 4.6 *Sense of rotation of a turntable as seen from above and below.*

Imagine yourself on a very large merry-go-round whose rate and sense of rotation can be varied. This is the situation of an earthbound observer (Figure 4.7). At the North Pole, an observer is spinning through the vertical axis at a rate of one rotation per day in a counterclockwise sense. At the South Pole, an observer is also spinning through the vertical axis at the rate of one rotation per day, but in the opposite (clockwise) sense. At the equator, the observer is not spinning at all around his vertical axis, although his "merry-go-round" is turning end-over-end at the rate of one complete turn per day. At some latitude intermediate between the pole and the equator, the rate of rotation around the vertical axis is something between the one rotation per day at the pole and the zero rotation per day at the equator. It is the rotation around the vertical axis which concerns us most, because it is what affects the horizontal motion and causes air to deviate from a high-to-low pressure path. If the latitude is designated by ϕ and the earth's rate of rotation (1 rotation/day) is designated as ω, the rate of rotation around the vertical axis can be shown to be $\omega \sin \phi$.

Figure 4.8 shows what happens if you try to throw a ball from a rotating merry-go-round at some target riding near the edge of the platform. To the observer on the merry-go-round, it will appear that some force has caused the ball to curve to the right of the intended path in the case of counterclockwise rotation (to the left for clockwise rotation). An observer not on the merry-go-round would say that the ball moved along a straight line but that the target turned. The observer on the platform could account for the deflection of the ball from the target by supposing a deviating force. The same thing happens on the rotating earth—the line connecting the target (a low pressure area) and a parcel of air is continuously changing its orientation.

The fictitious deviating force (fictitious *only* as far as a nonrotating observer is concerned) introduced to account for the effect of earth rotation is known as the

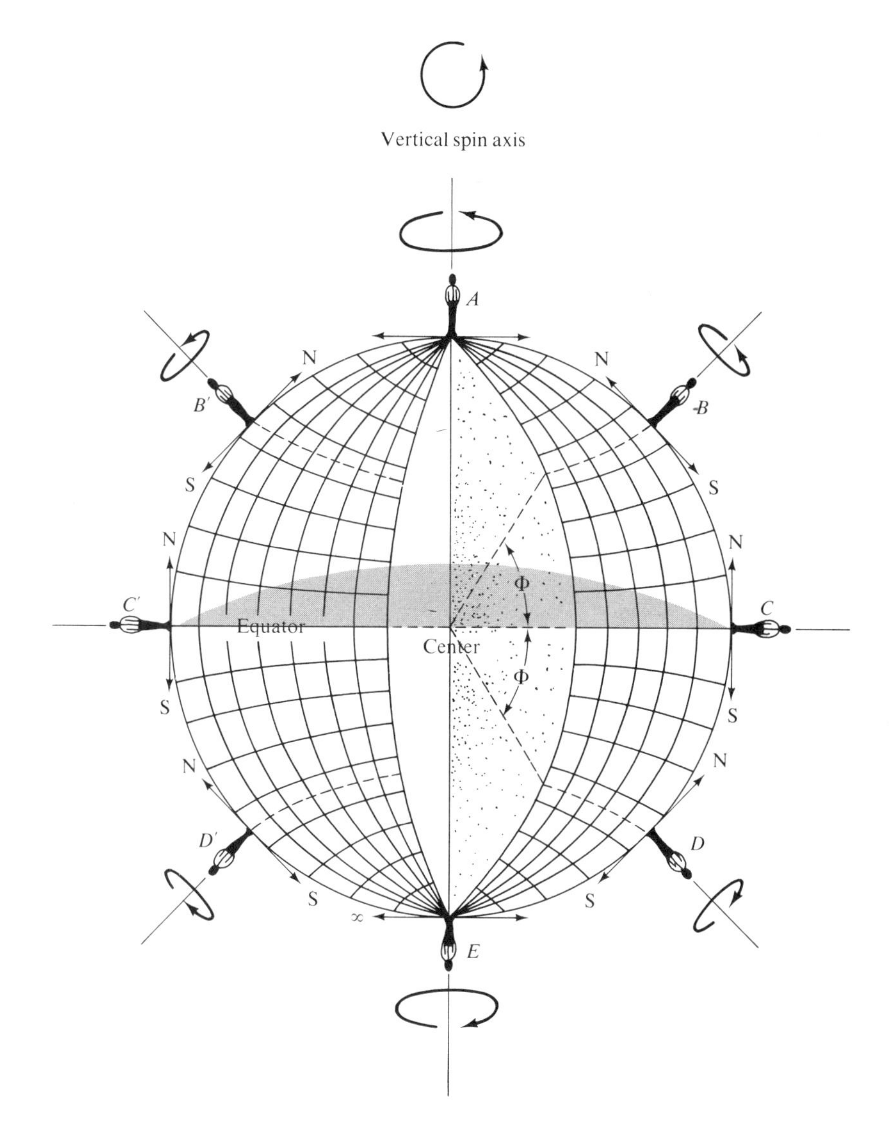

FIGURE 4.7 *Horizon rotation rate as a function of latitude.*

Coriolis force, named for the French mathematician who first explained it mathematically. The algebraic expression for the apparent acceleration is $2V \omega \sin \phi$, where V is the speed of the particle relative to the earth's surface. The magnitude of the acceleration is thus not only dependent on the latitude (maximum at the poles, zero at the equator), but also it is directly proportional to the speed of the particle. It always acts at a 90-degree angle to the wind. Although the Coriolis force is small, it is significant in horizontal air flow because, first of all, the horizontal pressure gradient force is also relatively feeble and, second, the air traverses great distances. It is very important in the large-scale flow, such as that associated with systems that affect the weather over thousands of miles, but it is of much lesser conse-

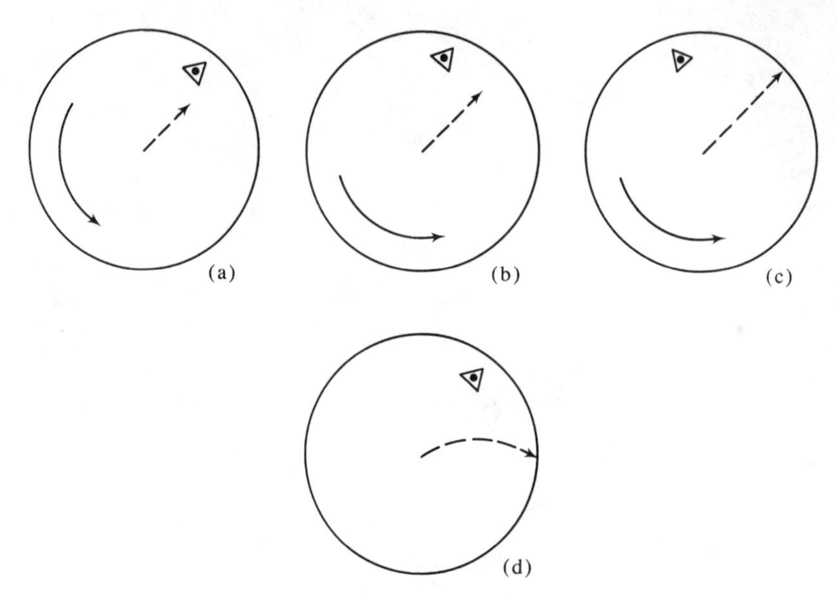

FIGURE 4.8 *Effect of rotation on apparent path of a moving body. A projectile is fired from the center of a merry-go-round at a triangular target (a). As the target rotates (b–c), the projectile misses the target to the right. To an observer rotating on the merry-go-round the projectile appears to curve to the right (d).*

quence in purely local, small–scale winds. The same, of course, is true in its effect on any body moving freely over the earth's surface. The correction must be applied to long–range artillery, but it has no detectable effect on the path of a bullet. Ocean currents are appreciably affected.

The Coriolis force comes into effect as soon as a particle has motion. If, for example, a particle had a westward velocity in the Northern Hemisphere and there

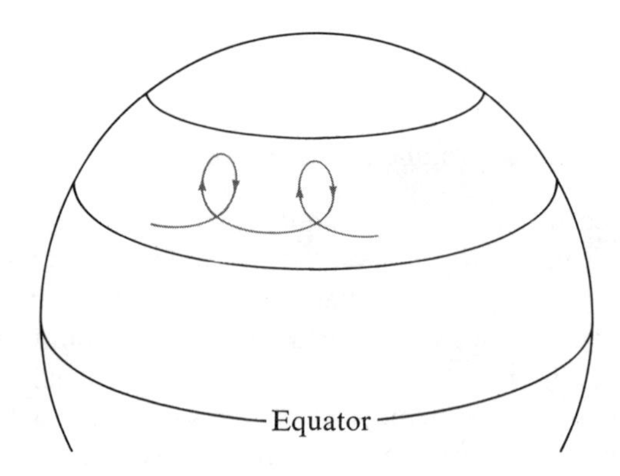

FIGURE 4.9 *Path of an unaccelerated moving particle relative to the earth's surface.*

were no forces acting on it other than the Coriolis force, it would take a path relative to the earth's surface like that shown in Figure 4.9. The form of the path would depend only on the speed of the particle and the latitude (rotation speed of the earth's horizon). For all moving parcels of air away from the equator, the Coriolis force constantly deflects the air to the right (in the Northern Hemisphere) or to the left (in the Southern Hemisphere). We can think of this apparent force as real, since we are concerned with the air motion relative to the earth's surface.

At elevations more than a kilometer or so above the ground, frictional effects are small, and a near balance exists between the horizontal pressure gradient, directed toward low pressure on the left of the direction of motion, and the Coriolis force, directed toward the right of the motion (Figure 4.10). This important balance is called *geostrophic equilibrium,* and a wind in perfect balance with the horizontal pressure gradient force is called the *geostrophic wind.* Mathematically the geostrophic balance is given by

$$fV_g = \frac{1}{\rho}\frac{\Delta P}{\Delta n} \tag{4.4}$$

where $f = 2\omega \sin \phi$ and $\Delta P/\Delta n$ is the magnitude of the horizontal pressure variation measured perpendicular to the flow. For example, the value of ω is $7.27 \times 10^{-5}\text{s}^{-1}$ so that at 43°N f is $1.0 \times 10^{-4}\ \text{s}^{-1}$. For the pressure gradient shown in Figure 4.10 of 4 mb/400 km (1×10^{-3} N/m^3) and a density ρ of 1.2 kg/m^3, the geostrophic wind

$$V_g = \frac{1.0 \times 10^{-3}\ \text{N/m}^3}{(1.0 \times 10^{-4}\ \text{s}^{-1})\ (1.2\ \text{kg/m}^3)} = 8.3\ \text{m/s}.$$

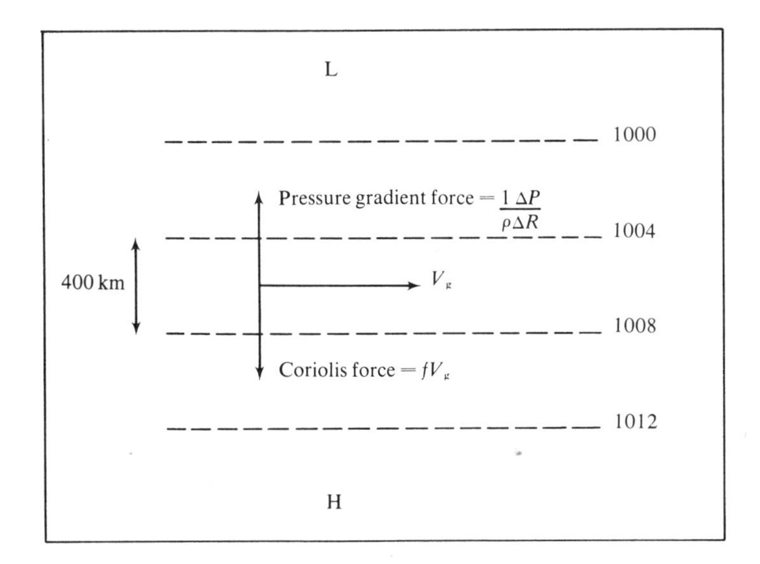

FIGURE 4.10 *Geostrophic equilibrium in which the horizontal pressure gradient force is balanced by the Coriolis force.*

Friction

Everyone is familiar with the fact that if a wooden box is given a push along a level floor, it will travel a short distance and then stop. The force that retards the forward motion is friction. It is the result of the interlocking of surface irregularities and the adhesion of touching molecules of the two contacting surfaces.

Although the adhesion is much less and the space between molecules is much greater in a gas, there is, nevertheless, a frictional drag created when velocity differences arise within a gas. This retardation of motion in a fluid is referred to as *viscosity.* When only the random, thermal motion of the *molecules* is responsible for this slowing up, the retardation, sometimes called *molecular viscosity,* is quite low. The effect of molecular agitation can be explained as follows: if a stream of air is directed along a solid surface, the air molecules in contact with the surface will have no motion (other than the usual random agitation) because they are blocked by the stationary molecules of the surface. These molecules will, in turn, retard the flow of those molecules adjacent to them because there is a continuous exchange of zero-velocity "surface" molecules with those in the next tier. Some of the slow-moving molecules of the second tier mix with those of the third tier, and so on, causing a progressively lesser retardation with distance from the surface.

The molecular viscosity of air is so small that if it alone were responsible for frictional drag in the atmosphere, the slowing up of the air flow would almost completely disappear within a meter of the surface. Far more significant is the so-called *eddy viscosity* which, at least in the lower layers of the atmosphere, is about 10,000 times more effective than molecular viscosity. As the name implies, it acts through the transfer of momentum between layers of air by *eddies* rather than by molecules.

Eddies are merely chunks of air that leave their places within otherwise orderly, smooth–path flow. A wind record marks the passage of these eddies as rapid, irregular fluctuations in direction and speed. Figure 1.6 shows these fluctuations as recorded by an anemograph. Such fluctuations—deviations from the mean velocity—are referred to as *turbulence.* In addition to momentum, turbulent fluctuations greatly accelerate the transport of the other quantities in a fluid. The most visible of these are pollutants, such as smoke. As a puff of smoke is carried along by the mean wind, eddies gradually diffuse the smoke particles over a bigger volume until the density of particles is so small that the puff can no longer be seen. Turbulent diffusion also spreads out the moisture and heat of the atmosphere.

The intensity of turbulence in the atmosphere depends on several factors, but the most important is the stability of the atmosphere—how well the atmosphere arrests vertical displacements of air parcels (see Vertical Stability), the roughness of the ground, and the speed of the wind. Figure 4.11 illustrates types of behavior of a smoke plume under three sets of atmospheric conditions.

The effect of turbulence on the wind is to cause a transfer of momentum through a much deeper layer of air than would occur if only molecular diffusion processes were operating. Depending on the speed of the air flow, the roughness of the underlying surface, and the stability of the atmosphere, the drag of the surface

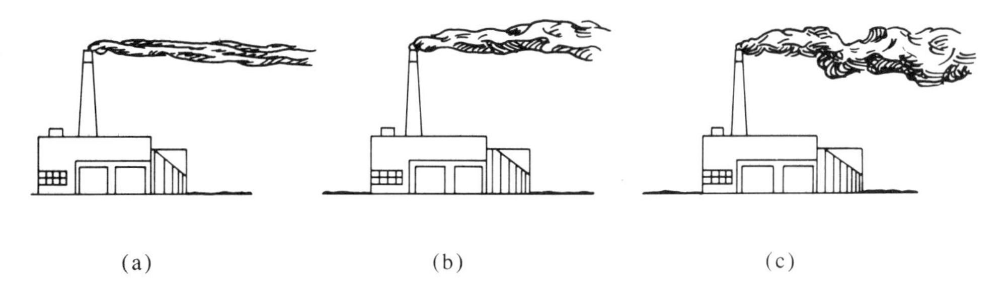

FIGURE 4.11 *Effect of turbulence in diffusing smoke: (a) laminar (non-turbulent) flow; (b) partially turbulent flow; (c) well-developed turbulence.*

on the flow can extend from anywhere between a 300-meter and a 2000-meter elevation. On a day when the atmosphere is well-mixed [curve *A* in Figure 4.12(*a*)], the surface can be "felt" as high as 2000 meters. In contrast, when the atmosphere is stratified and vertical mixing is suppressed [curve *B* in Figure 4.12 (*a*)], the surface drag extends upward to only 300 meters or less. The speed of the wind at the anemometer level (usually 5–10 meters above the surface) is only a fraction of the speed in the free atmosphere.

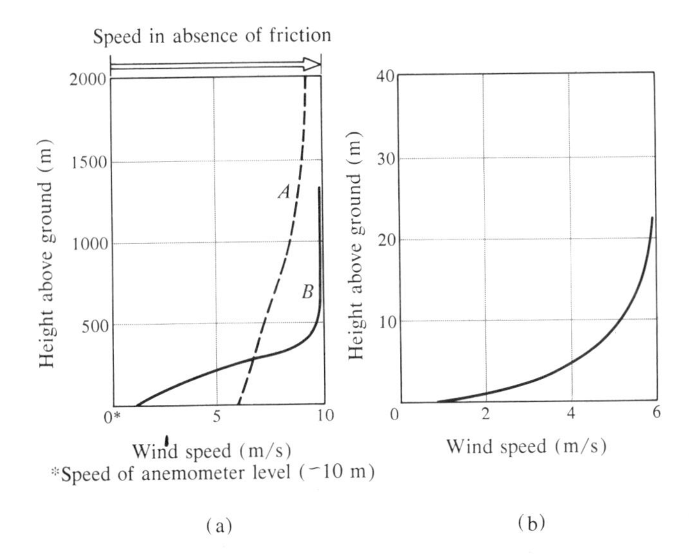

FIGURE 4.12 *Examples of the variation of wind speed with height in the surface "friction layer"; (a) A = strong vertical mixing; B = weak vertical mixing; (b) variation of wind speed in the first 40 meters.*

The effect of friction on the horizontal wind that is observed near the ground is illustrated in Figure 4.13. Near the surface, the frictional drag (always opposed to the direction of air motion) slows the wind; the Coriolis force is correspondingly less, and the "steady," or mean, wind is that resulting from a balance of three forces—

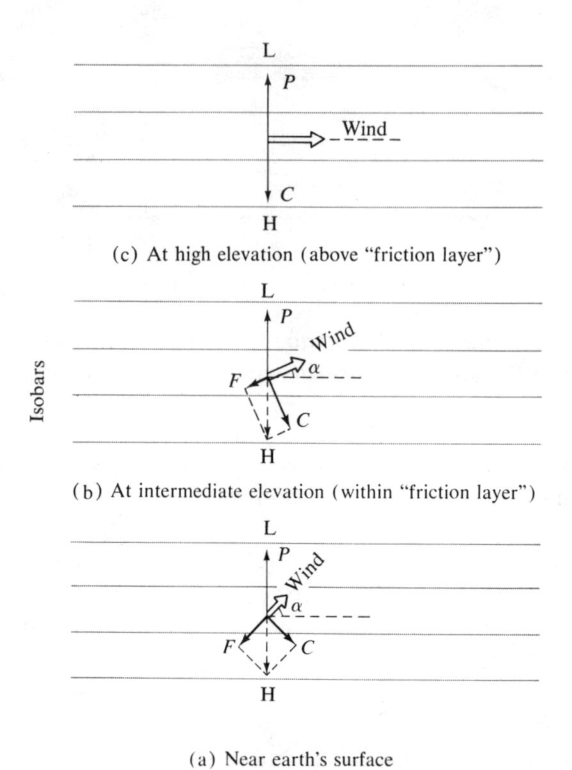

FIGURE 4.13 *Wind velocity variation in the "friction layer." (Northern Hemisphere, constant pressure gradient, α is the angle between the wind direction and the isobars.)*

pressure gradient, Coriolis, and friction. Since the Coriolis force and friction are always at 90 degrees and 180 degrees from the wind direction, respectively, a balance of the three forces can be achieved only if the wind blows at an angle across the isobars. The angle can be 45 degrees or more, but it is usually only 20 degrees or 30 degrees. At higher elevations, the wind speed increases and the direction cuts across the isobars at a smaller angle; the three forces are again in balance as far as the steady wind is concerned. Above the friction level (i.e., where friction becomes negligible) the wind is stronger still and its direction is essentially parallel to the isobars. As mentioned earlier, such a wind, resulting when the only forces acting are the Coriolis and pressure gradient, is called *geostrophic*. The level at which the wind is a close approximation of the geostrophic is normally above 300 meters.

The above discussion of the variation of wind with height and Figure 4.13 assume that the horizontal pressure gradient does not vary with height. If the horizontal pressure gradient changes with increased height, the wind will change due to this effect as well as to the decreased frictional drag of the ground. In fact, the horizontal pressure gradient normally does vary greatly with altitude. In middle latitudes the pressure gradient force normally reaches a maximum around 8 km, and it is at this level that the fastest winds (jet stream) occur.

Curved paths

According to Newton's first law of motion, the velocity of a body does not change as long as there is no net (unbalanced) force acting on it. But keep in mind that velocity is a vector; i.e., it has both magnitude (speed) and direction. Thus, an object that moves at constant speed along a curved path is changing direction and is, therefore, experiencing an acceleration. This acceleration is directed at 90 degrees to the path, inward along the turning radius. It is called centripetal acceleration.

A centripetal force must be applied to make an object deviate from its natural tendency to move along a straight line. For example, the muscles of a discus thrower apply the centripetal force needed to make the discus move on a circular path until the moment that he releases it; then, it moves along a straight line. We can think of this centripetal force that is making the object follow a curved path as opposed to a "centrifugal force" that is apparently pulling the object outward from the desired path.

Air parcels experience a centripetal acceleration whenever they travel along paths that are curved relative to the earth's surface; they are forced to follow such a path whenever the isobars are curved or circular. The magnitude of this acceleration is given by V^2/r, where V is the wind speed and r is the radius of curvature of the air parcel's path. Thus, the tighter the curve and the faster the wind speed, the greater is the centripetal force required. In air flow having gentle curvature, such as that in a circulation pattern that covers half a continent or more (see next chapter), the centripetal force is negligible compared to other forces. But in the case of a small, intense whirlpool, such as a tornado, in which r may be only 100 meters or less, the centripetal force must be very large.

The types of flow patterns associated with low pressure areas (*cyclones*) and high pressure areas (*anticyclones*) near the earth's surface in the Northern and Southern hemispheres are shown in Figure 4.14. In the Northern Hemisphere, the air is seen to spiral in a counterclockwise sense into a cyclone; it spirals in a clockwise sense out of an anticyclone. In the Southern Hemisphere, the spiral is inward in a clockwise sense for a cyclone; it is outward in a counterclockwise sense for an anticyclone. The approximate locations of high and low pressures can be determined from the observed wind direction at a point through *Buys Ballot's rule:* If you stand with your back to the wind, low pressure is to your left, high pressure is to your right (Northern Hemisphere).

4.3 Vertical Motion and its Relation to Clouds

We have seen how differences in temperature over the globe lead to the creation of thermal circulations, which involve both vertical and horizontal motions. Because of the important role played by vertical motion in producing weather, some further discussion of how vertical motion is produced and how it affects the formation and dissipation of clouds will be given here.

In discussing the characteristics of the troposphere (Chapter 1), it was mentioned that convection keeps this lowest layer fairly well stirred, in contrast to the

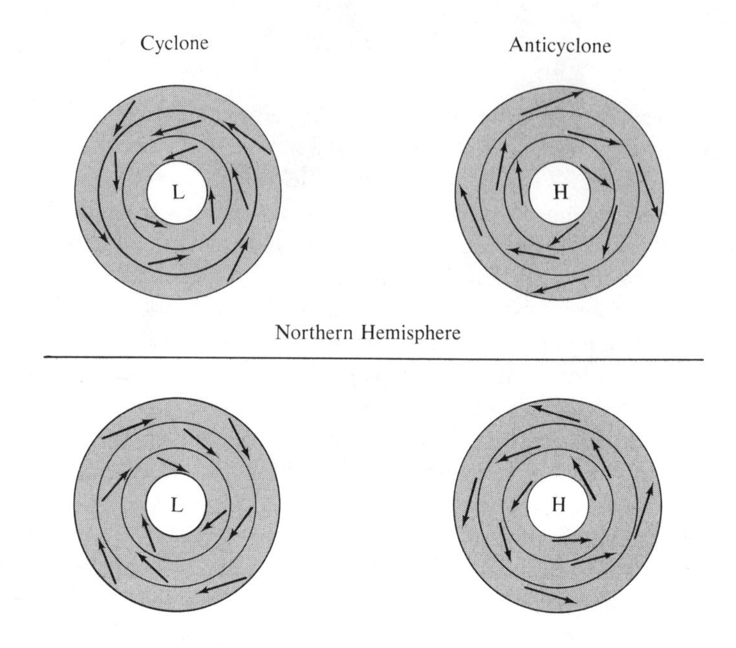

FIGURE 4.14 *Cyclonic and anticyclonic flow near the earth's surface.*

stratosphere, in which there is not very much mixing. Yet, on the average, the temperature in the troposphere is not uniform in the vertical, but rather it decreases at the average rate of 6½ °C/km. Evidently, a well–mixed layer of air does not imply one of constant temperature, at least not in the vertical. The reason for this is linked to the pressure changes that air parcels experience during displacements.

Adiabatic processes

Air displaced vertically experiences especially rapid pressure changes; in response to these changes, the volume must increase or decrease. For example, if a parcel of air is forced to descend from an elevation of one kilometer, the pressure exerted on it will have increased by about 20 percent by the time it reaches the surface. In response to such a pressure change, the volume and/or the temperature must also change. An indication of what actually occurs can be obtained from our experience in letting air escape from a tire—as the air expands and the pressure drops rapidly, the air cools. The air cools because some of its heat energy is expended in doing work of expansion. Conversely, if we pump up the tire rapidly, the air warms up because the work we have done in compressing the gas is converted to heat. If we neglect the slow dispersion of heat by conduction through the walls of the tire, we could compute the temperature change by equating the work done in expansion or compression to the change in heat content of the air.

The same process takes place when the pressure of an air parcel is rapidly changed by its descent or ascent in the atmosphere. Assuming that the heat loss or gain by an air parcel through conduction, radiation, and mixing with the surroundings is at a slow enough rate during a vertical displacement, the temperature changes can be ascribed mostly to volume changes.

The ideal or theoretical process during which there is absolutely no heat exchange between a mass and its environment is said to be *adiabatic*. Since air usually contains water, and phase changes involve latent heat, we distinguish between two different adiabatic processes: (1) A *dry adiabatic* process is one during which there are no phase changes of water (no condensation, evaporation, fusion, or sublimation). (2) A *moist* or *wet adiabatic* process is one during which phase changes *do* occur, and account must be taken of the latent heat.

Since the horizontal pressure gradient is small, and the wind crosses the isobars at a small angle, compression or expansion of air parcels moving in the horizontal is very small. For this reason, we are concerned only with volume changes of air parcels when they move up or down. A dry adiabatic displacement in the vertical results in a temperature change of about 1°C for every 100 meters of elevation (5½°F/1000 ft). A parcel of air moving upward from near sea level (pressure ~ 1013 millibars) in a dry adiabatic process to an altitude of 7000 meters (pressure ~ 410 millibars) would almost double its volume and its temperature would drop almost 70°C (Figure 4.15). If the same piece of air were to be returned to its original level, its temperature and volume would assume their initial values.

During a moist adiabatic process, the changes of phase of the water contained within a parcel of air experiencing a rapid pressure change cause conversion of latent heat to sensible heat, and vice versa. In other words, when condensation is occurring, the latent heat released raises the temperature of the air parcel; and when evaporation is occurring, the latent heat required cools the air. In the example shown in Figure 4.15, if the air parcel at sea level were saturated with water vapor, each kilogram of dry air would contain about 10.7 grams of water vapor; any lifting of the parcel would cause expansion and cooling, and the excess moisture would have to condense. Thus, when the parcel reached an altitude of 2000 meters (pressure ~ 800 millibars), the air would still be saturated with water vapor mass, but the amount of moisture in vapor form would be less than two thirds the original value. More than 4 grams of water in each kilogram of air would have condensed, releasing about 4 × 600 = 2400 calories of latent heat. As a result, the temperature of the air parcel would be considerably warmer (9°C) than it would have been had the process been dry adiabatic. Further ascent of the parcel would cause more condensation, but because the rate at which condensation proceeds is less when the temperature is low than when it is high, the rate of temperature change during a moist adiabatic process is not constant. In Figure 4.15, this can be seen by comparing the temperature change between 2000 meters and 4000 meters (about 6°C/km) with that between 4000 meters and 7000 meters (about 7.3°C/km).

If the "moist" air parcel illustrated in Figure 4.15 were to descend, it would warm at the same rate that it cooled during the ascent, *if* all of the liquid and solid particles remained in the parcel. In practice, some of the water drops and ice crystals leave the parcel of air, so that if the parcel later descends, there will be less evaporation and hence a more rapid warming than occurs in the true moist adiabatic process. When some of the condensation products drop out, the process is said to be pseudo-adiabatic because it is "irreversible."

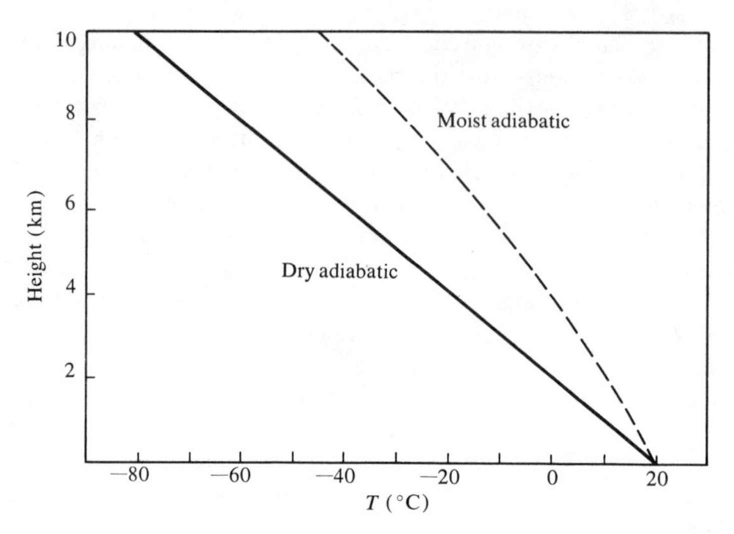

FIGURE 4.15 *Variation of temperature in a parcel of air as it is lifted dry and moist adiabatically from a height of 1 to 10 km. The dry adiabatic lapse rate is a constant 9.8°C/km; the moist adiabatic rate varies from about 6°C/km near the surface to 9.8°C/km at 10 km.*

Condensation in the atmosphere is produced principally through the cooling of air as it ascends, comes under lower pressure, and expands. Similarly, the dissipation of clouds is usually a sign of descending air. The overall pattern of vertical motion producing a cumulus cloud, such as that shown in Figure 4.18, is upward motion below and within the cloud and downward motion at the edges outside the cloud. (In a well-developed cumulonimbus cloud, the pattern is more complex, as we shall see later.) Even air containing little water vapor (low relative humidity) does not require a great deal of vertical lift to create saturation and then condensation. For example, air starting near sea level with a temperature of 30°C and a dew point of 14°C (relative humidity = 36 percent), will become saturated at about 2000 meters elevation. (The dew point decreases at the rate of almost 2°C per 1000 meters of ascent during a dry adiabatic process. The height at which saturation will be reached is, therefore,

$$\frac{(30° - 14°)}{1°/100\text{ m} - 0.2°/100\text{ m}} = 2000\text{ m} \tag{4.5}$$

After saturation, both the dew point and temperature decrease at the moist adiabatic rate.)

4.4 Vertical Stability

The ability of the atmosphere to produce and sustain vertical currents depends on the atmosphere's "stability." A *stable* atmosphere is one in which buoyancy forces oppose the vertical displacement of air parcels from their original levels. An *unstable* condition exists when buoyancy forces abet the vertical displacement of air parcels. A *neutral* state exists when vertical displacement is neither opposed nor abetted by buoyancy forces.

The buoyancy of a parcel of air will depend on its density relative to the environment density at the same level. If a parcel is "heavier" than the medium surrounding it at the same level, it will be forced to sink; if it is lighter, it will be forced to rise. If its density is the same as that of its surroundings, there will be no "Archimedean force" tending to make it either rise or fall.*

Since we do not normally measure density directly in the atmosphere, it is more convenient to discuss stability in terms of temperature, rather than density. From the Equation of State (p. 10), we know that at any fixed pressure the density is inversely proportional to the temperature. Therefore, at any given pressure level, we can substitute temperature for density in the statements on buoyancy in the previous paragraph: a parcel of air that is warmer than its surroundings will tend to be pushed upward; one that is colder than its surroundings, downward; and one at the same temperature as its environment will not experience a "push" in either direction.

We have seen from the discussion of adiabatic processes that the temperature of an air parcel displaced vertically changes at a *fixed* rate – 1 °C/100 meter if there is no water phase transition and at some lesser rate if there is condensation or evaporation. Evidently, then, whether a vertically displaced parcel of air is warmer or colder than its environment at any particular point along its path will depend on the vertical distribution of ambient atmospheric temperature. The rate at which the temperature decreases vertically in the atmosphere is called the *lapse rate*. (For example, the *average* lapse rate in the troposphere is 6½ °C/km.)

An atmospheric layer in which the lapse rate is less than the adiabatic is stable. A stable atmosphere is indicated by temperature sounding (profile) *A* in Figure 4.16. If a parcel of air at 500 meters is displaced upward, its temperature will decrease at the dry adiabatic lapse rate of 9.8 °C/km and will follow the solid curve of Figure 4.16. Thus it will quickly become colder than its environment at the same level and buoyancy will force the parcel back down. Or, if the same parcel is displaced downward, it will always be warmer than the environment, so that buoyancy will force it back up. In other words, this is a *stable* temperature lapse rate because vertical motions are suppressed. In the *unstable* case (sounding *B* in Figure 4.16), a parcel displaced upward will be warmer than the environment at each level, and if it is displaced below the reference level, it will be colder than the environment. Thus, the lapse rate is such that vertical motions are abetted. In the *neutral* case, the lapse rate is exactly equal to the dry adiabatic rate, and therefore the temperature of the parcel at any new level is equal to that of the surrounding air. The criterion, then, for stability in the case of *dry adiabatic* processes is that the lapse rate in the atmosphere layer be less than 1 °C/100 meter; for instability, greater than 1 °C/100 meter; and for a neutral state, equal to 1 °C/100 meter.

Of course, ascending air currents very quickly become saturated, after which the air cools at the *moist adiabatic* rate. The lapse rate criteria would then be based

*Archimedes (born 287 B.C.) used the principle of buoyancy to determine the percent of gold in the crown of Hiero, King of Syracuse.

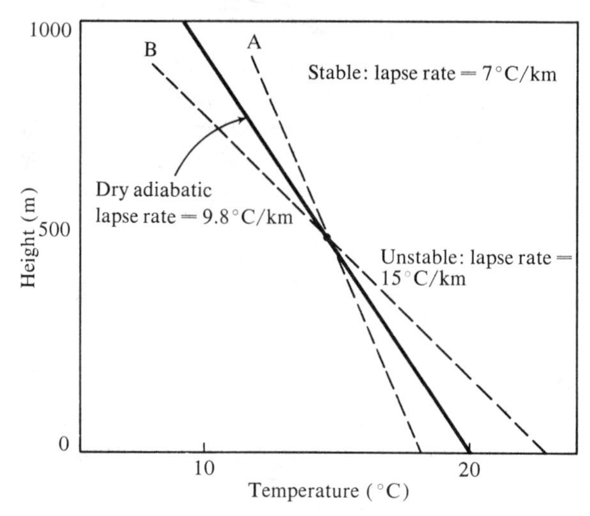

FIGURE 4.16 *Stable (A) and unstable (B) temperature profiles.*

on the moist, rather than the dry, adiabatic rate. For example, the stable situation shown in Figure 4.16 would be *unstable* for upward displacement if the air parcel were saturated. In ascending, the parcel would cool at the rate of 6°C/km, so it would always be *warmer* than its environment. Thus, condensation of water with its release of latent heat is an important factor in inducing vertical currents of air. We see here the basis for an earlier statement that a significant portion of the energy that drives the atmosphere is brought in through the evaporation-condensation cycle of water.

Changes in stability

Lapse rates vary considerably in space and time. Increase of the lapse rate in a layer of atmosphere results from warming of the lower part of the layer and/or cooling of the upper part. Conversely, a decrease of the lapse rate is produced by cooling of the lower portion and/or warming aloft. Some of the possible causes for different rates of temperature change in a layer follow.

(1) Differential advection If the air aloft is being replaced by warmer air brought in by the winds, while in the lower portions colder air is being brought in, the stability of the layer will increase. The opposite, cooler air coming in at high levels with warmer air near the bottom of the layer will increase the instability of the layer.

(2) Surface heating or cooling This can occur in either of two ways: (*a*) It occurs when air moves over a surface that is either colder or warmer than itself. For example, air moving from an ocean over a warm continent may cause enough instability to set off showers. In the winter, warm air from the Gulf of Mexico streaming northward over the central and eastern United States sometimes is cooled sufficiently by the cold surface to produce widespread areas of fog and low layers of stratus clouds. (*b*) When air is over a surface that is losing or gaining heat through radiation the air in contact with the surface will cool or warm. The cooling of the air near the surface on a clear, calm night frequently leads to the creation of a temperature inversion (i.e., a layer of *increasing* temperature with height).

(3) Radiative cooling aloft Clouds are quite effective blankets, retaining the heat of the air below them. The loss of heat at the *tops* of clouds, though, can lead to greater instability within the clouds.

(4) Vertical displacements of layers When an entire layer of air sinks (*subsidence*), the difference in the percent compression between the bottom and top of the layer leads to a greater warming of the upper portion than of the lower portion of the layer, and therefore greater stability. Inversions created in this manner are known as *subsidence inversions*. Conversely, when a layer of air is lifted *dry* adiabatically, its stability decreases. However, if part of the layer becomes saturated during the ascent, the situation changes. If the upper portion becomes saturated before the lower, the stability will be increased, because the upper zone will cool at a lesser rate during the ascent. But if the lower portion becomes saturated earlier, because the upper part will cool faster than the lower, the instability within the layer will be enhanced. Thus, instability is more likely to occur in a layer of air if the bottom of the layer is more nearly saturated at the onset of lifting.

Stability and clouds

Cumuliform clouds are associated with instability and strong vertical motion. The convective ascent of air in such clouds appears to occur in bursts or bubbles, somewhat like those that form in boiling water. Each successive bubble, having dimensions of up to a few kilometers in the horizontal and a few hundred meters in the vertical, rises, expands, and cools; but, as the bubble ascends through the atmosphere it is "eroded" or mixed with the surroundings, gradually losing its identity (Figure 4.17). In this

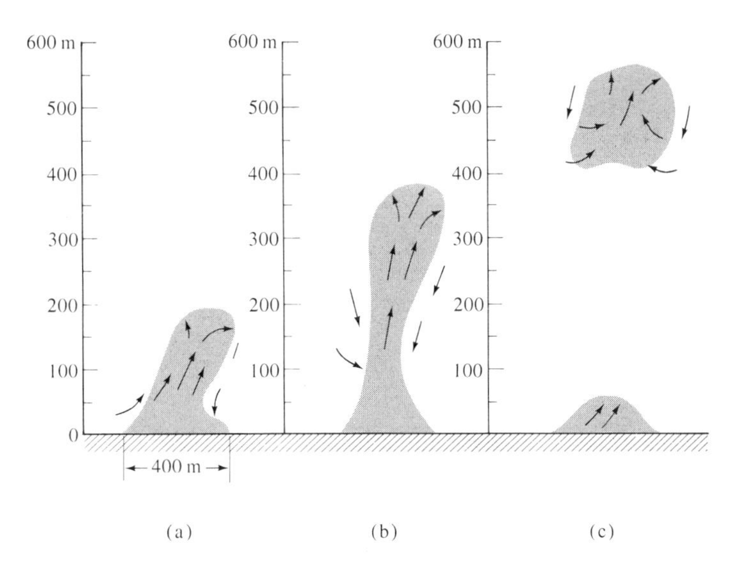

FIGURE 4.17 *Development of convective bubbles producing a "thermal." (Note that the vertical scale is exaggerated.)*

way, puffs of cumulus may form, gradually disappear, and perhaps be replaced by new puffs. Horizontal winds may carry each bubble downstream from the surface point where it was created. Strong, turbulent flow tends to cause rapid erosion.

However, if the horizontal winds and turbulent mixing are not too great, previous bubbles may still remain while new bubbles are being formed. Then, each subsequent bubble will be able to ascend to ever greater heights, causing the cumulus cloud to develop vertically. Of course, an air bubble cannot rise without some other air descending to replace it, since there can be no vacuum. With increased instability, enhanced by the release of latent heat within the cloud, the "percolation" becomes more continuous and a well-defined thermal circulation, such as occurs in the towering cumulonimbus of a thunderstorm cell, may develop. The rapidity with which such a convective cell can develop is illustrated in the photographs of Figure 4.18.

Columns of rising air are sometimes called *thermals*. Glider pilots learn to seek out these thermals and then try to remain within their boundaries so they can be carried upward. All around the thermal there is a compensating downward flow of air.

4.5 Causes of Vertical Motion

Vertical displacements of air result from "dynamic" causes as well as from changes in static stability. One dynamic cause is mountains. They act as barriers to horizontal air flow, forcing it to ascend along the windward sides and descend on the lee sides. On a large scale, another important cause of vertical motion is the divergence and convergence of air currents circulating around the great anticyclonic and cyclonic whirls of the atmosphere. Air currents spiraling inward at low levels of cyclones converge; i.e., they move toward the center. Since the horizontal area occupied by a volume of air must therefore decrease with time, the vertical depth must increase. This is illustrated by Figure 4.19(a). Imagine a column of air, having the boundaries shown in the figure, with streams of air spiraling inward toward the center. The *inward* component of the flow would result in a shrinking of the cross-sectional area with time (*convergence*). Since the amount of mass contained in the imaginary cylinder must remain constant, it follows that the depth of the cylinder must increase. Air must therefore move upward within the column. Conversely, in the lowest few kilometers of anticyclones, outward flow everywhere would result in an expansion of the column's horizontal cross section (*divergence*); vertical shrinking would be required to keep the total volume constant. At levels above 6 or 7 kilometers, horizontal convergence occurs over surface anticyclones, while divergence occurs over surface cyclones, so that a pattern of motion such as that illustrated in Figure 4.19(b) emerges. Note that the vertical scale in this drawing is greatly exaggerated. The diameter of the typical anticyclone or cyclone is greater than 1000 kilometers; thus, the downward and upward flows are not nearly as steep as they appear in Figure 4.19(b). The vertical velocities produced by these large-scale patterns of divergence and convergence are usually not more than a few centimeters per second (1 kilometer per day). However, they are sufficient to set the weather "stage" over large areas. In the absence of other influences, the weather over areas dominated by cyclones tends to be that of widespread cloudiness and precipitation, while that over anticyclonic areas is frequently clear.

1356 MST

1406 MST

1416 MST

FIGURE 4.18 *Stages of development of a cumulonimbus over Arizona. (Courtesy of L. Battan, University of Arizona.)*

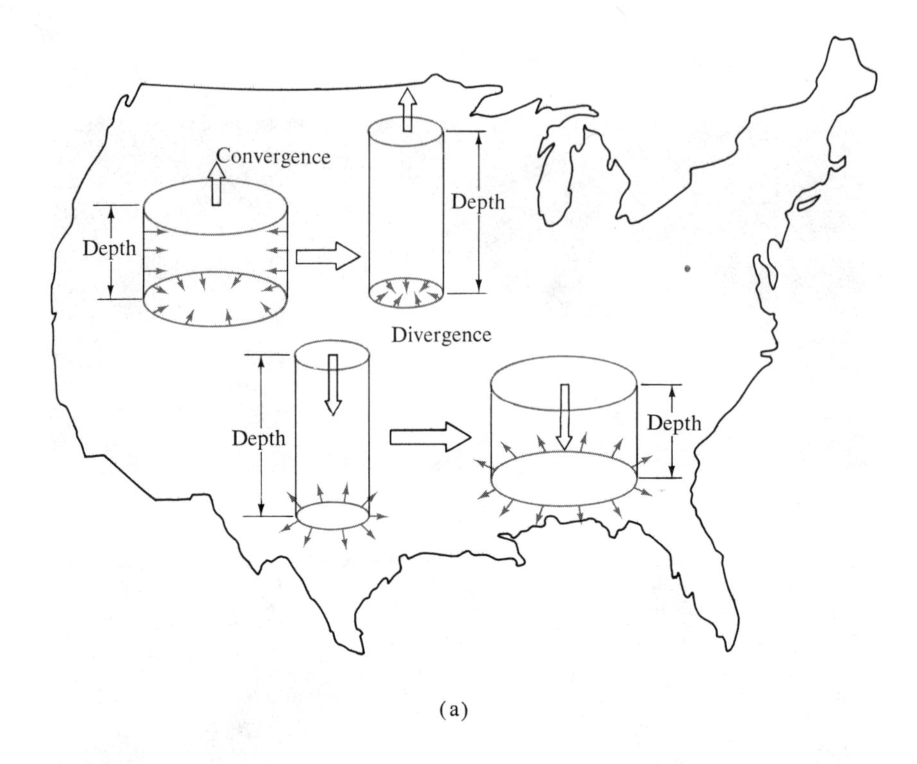

(a)

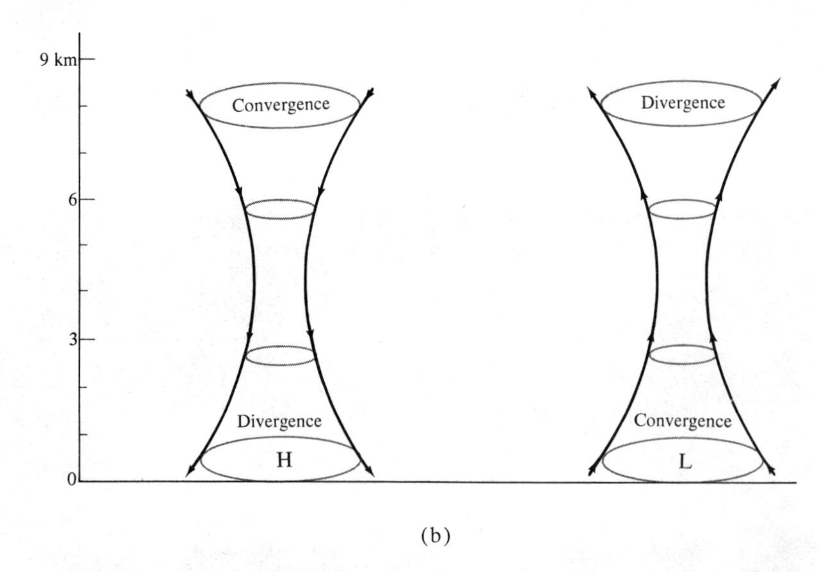

(b)

FIGURE 4.19 *(a) Convergence and divergence of a disk of air. (b) Large-scale patterns of divergence and convergence associated with surface anticyclones (highs) and cyclones (lows).*

PROBLEMS

1. In terms of density, explain how a balloonist is able to ascend or descend at will. Before light gases such as helium were available, balloonists inflated their balloons with hot air. What does this show about the effect of temperature on density and thus buoyancy? Why does smoke rise? Explain how portions of the atmosphere acquire buoyancy.
2. How would you classify the mean or standard lapse rate in the atmosphere as far as stability is concerned?
3. Plot a graph having as the abscissa temperature over the range of +30°C to −55°C, and as the ordinate, height over the range 0 to 16 kilometers. On the right-hand vertical scale, indicate the standard pressure in the vertical, as determined from Appendix 2 and Figure 1.2. Plot the following three temperature-pressure soundings measured in three different air masses by connecting consecutive points in each sounding with straight-line segments. Label each curve. Draw several straight, sloping lines on the chart to illustrate the rate at which temperature changes with height during a dry adiabatic process, one starting at sea level and 30°C, another at 0°C and sea level, and a third at −30°C and sea level.
   (a) Identify layers in the three soundings that exemplify absolutely stable, absolutely unstable, and neutral stratifications.
   (b) Identify all layers containing either an inversion or isothermal lapse rate.
   (c) Where would you say the tropopause is located in the first two soundings?

Air Mass	*Tropical*	*Polar (summer)*	*Polar (winter)*
Pressure (mb)	*Temp. (°C)*	*Temp. (°C)*	*Temp. (°C)*
1000	27	13	−31
950	22	13	−32
900	25	9	−32
850	22	4	−30
800	20	−1	−30
700	13	2	−28
600	6	−8	−31
500	−1	−17	−38
400	−13	−32	
300	−28	−50	
200	−50	−50	
150	−51	−50	

4. Compute the height of the base of a cumulus cloud formed by a thermal if the surface temperature and dew point are 85°F and 49°F, respectively.
5. If the earth were not rotating, what would be the direction of the horizontal wind with respect to the isobars? What force would keep the air from increasing its speed ad infinitum?
6. Explain the behavior of a Foucalt pendulum. At what rate does it rotate? (Review the discussion of the Coriolis force.)

7. Can there be a geostrophic wind at the equator? Explain your answer.
8. Explain the reasons for the horizontal and vertical pressure gradients in terms of density variations within the atmosphere.

Atmospheric Circulations

5.1 Scales of Motion

In the last chapter, it was shown that temperature differences produce the basic force that drives the winds. However, the picture of the air flow is made complex by the earth's rotation, frictional drag and turbulence, mountain obstacles, and, perhaps most of all, the extremely variable character of the earth's surface and the incessant changes of the state of water in the air. To simplify the analysis of the enormously complex patterns of "eddies within eddies" that exist in the atmosphere, it is convenient to categorize circulation systems according to size. Almost every size is represented in the atmosphere—everything from the very small whirls that kick up the dust on a road to enormous oscillations that have horizontal dimensions of several thousand kilometers. All of these different sizes—or *scales of motion,* as they are called—are interdependent. For example, an eddy produced by a hill might not occur unless there were a prevailing wind due to a circulation of much larger size.

An instantaneous snapshot of the winds of the entire atmosphere would present an extremely chaotic view of the flow. The complex distribution of forces producing such flow would make prediction an impossible task. To achieve some order, a type of filtering, according to size of flow elements, must be applied. This is accomplished by a system of averaging.

The very small-scale eddies or whirls that cause a wind vane to oscillate rapidly or branches of a bush to sway with periods of perhaps only a few seconds can be eliminated by averaging the observed wind velocity over periods of several minutes. If one were to average the wind velocity over an entire day, then wind oscillations having periods of much less than a day would disappear from the record. Meteorological observations are averaged over time and space to isolate the various sizes of atmospheric motions. The analyst of the weather maps that are published in the newspapers applies an averaging process—a smoothing of isobars —that eliminates most irregularities smaller than about 100 kilometers.

Actually, most routine meterological measurements are made in such a way that very small eddies are eliminated. Most anemometers and thermometers do not react to small, high-frequency changes. Observations are so widely spaced in time and area that most must be considered averages over horizontal distances of tens of kilometers and vertical distances of tens of meters. Even such relatively large circulation phenomena as thunderstorms and tornadoes often fall through the "mesh" of the usual weather station network.

The scales of atmospheric motions can be classified as shown in Figure 5.1. On the microscale are small, short-lived eddies, often referred to as turbulence, that are strongly affected by local conditions of both terrain roughness and temperature. These eddies are very significant as dispersers of pollutants. The Coriolis force is not significant on the microscale. At the large end of the microscale are tornadoes and waterspouts.

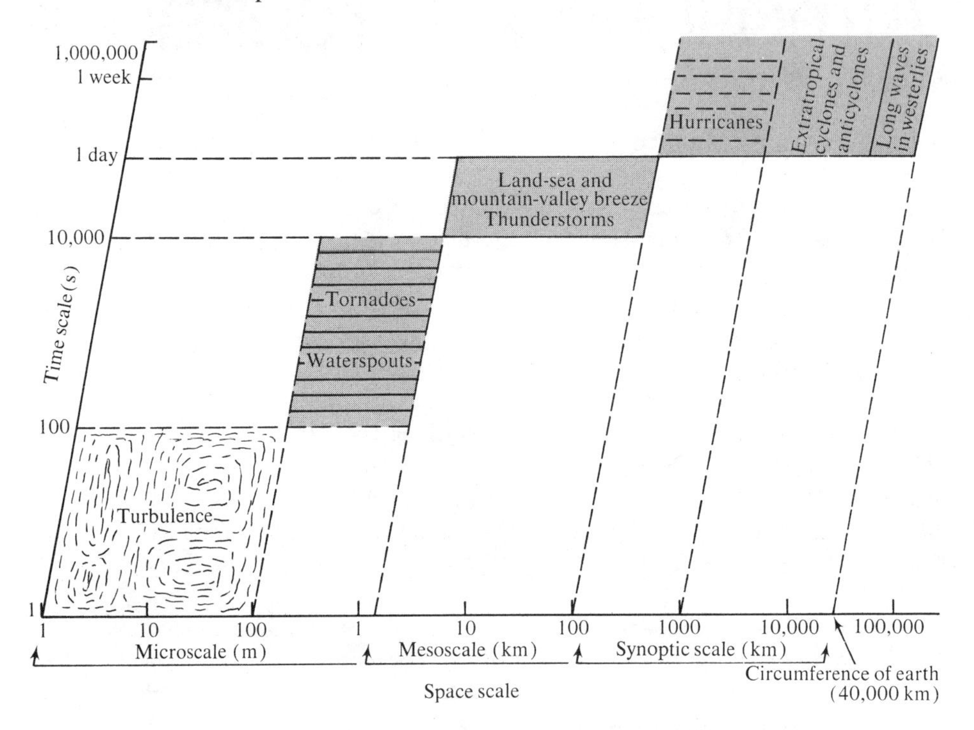

FIGURE 5.1 *Horizontal and temporal scales of atmospheric circulations.*

The mesoscale includes a variety of phenomena of intermediate horizontal size such as land-sea breezes, thunderstorms, and squall lines. Coriolis forces may play an important role in the larger phenomena in this class, such as sea breezes.

The cyclones and anticyclones that are largely responsible for the day-to-day weather changes belong to the large or synoptic* scale. These systems persist for days or weeks and Coriolis forces are very significant. At the largest end of the synoptic scale, sometimes called the planetary scale, are features of the atmospheric circulation that persist for weeks or months. Long waves that exist in this flow move very slowly, or not at all, across the earth. These play an important role in determining the seasonal characteristics of the weather. Coriolis forces are very important on this scale.

*In meteorology, *synoptic* means "coincident in time." Thus a synoptic weather map shows conditions as they were at a particular time. The word "synoptic" has come to denote large-scale as opposed to small-scale atmospheric patterns.

The wind observed at any place can then be thought of as a composite of several different scales of motion. For example, the synoptic flow patterns are associated with the large features of the earth's surface and distribution of heat: continents and oceans, extensive mountain ranges, latitudinal variations of insolation. The various scales of motion can also be characterized by the magnitude of the vertical motion associated with each. Macroscale motion is mostly in the horizontal. The vertical displacements attributable to the very large circulation features are no more than 1 or 2 cm/s (0.04–0.07 km/h); even in the great cyclonic storms that regularly affect the middle and high latitudes, average vertical displacements are only of the order of 50 cm/s (1.8 km/h). In the smaller, more intense mesoscale circulations, the vertical velocities are more comparable to the horizontal velocities; for example, in a thunderstorm, the vertical motion is often 10 m/s (36 km/h) and can reach 30 m/s or more. Winds of the microscale are generally much weaker than those of the larger sized motions, but the vertical motions are very nearly equal to those in the horizontal. However, microscale motions, unlike those of the mesoscale, appear to occur principally in a rather shallow layer adjacent to the earth's surface.

5.2 The Nature of Atmospheric Circulations

At the beginning of Chapter 3 attention was drawn to certain similarities between the heat engine and the atmosphere. This analogy can be demonstrated in yet another way. Just as the motor of an automobile functions through the turning of wheels and gears of various sizes, the total kinetic energy of the atmosphere is partitioned among circulations of varying dimensions. Furthermore, in the mechanical engine, the larger the mass of the rotating wheel or gear, the greater the power required to turn it. In the atmosphere, also, the amount of energy required to initiate and maintain a circulation pattern, is for the most part, directly proportional to the mass of air associated with the circulation.

The wind system which contains the greatest mass of air is called the *general circulation*. Driven by the energy received from the sun, it serves to transport the air from the equatorial regions toward the poles, and to maintain a return flow of cold air from polar to tropical latitudes. It determines, in large measure, the broad pattern of climates of the earth. Within this large-scale global flow are embedded the smaller circulations. These smaller circulations—"perturbations" on the worldwide flow—are responsible for the transient, short-period variations of atmospheric conditions, i.e., the weather. Here, we shall examine the structure and causes of various sizes of atmospheric circulation patterns.

The general circulation

The general circulation of the atmosphere is a time-averaged flow of air over the entire globe. It is determined by averaging wind observations over long periods of time—usually twenty years or more. In order to isolate the seasonal variation of the general circulation induced by the earth's revolution about the sun, the averaging is sometimes done separately for each season of the year. In any case, this long-period averaging tends to eliminate the smaller circulations, as we explained at the beginning of this chapter.

If the earth were not rotating and if the surface were homogeneous, solar heating at the equator would cause the air in that region to rise and flow toward the poles. As it was transported poleward, not only would the air become cooler and tend to sink toward the surface, but the convergence of the meridians of longitude would force the air to "pile up" before it reached the pole. These effects would induce a return circulation near the earth's surface from polar regions to the equator. In practice, however, this simple thermal circulation pattern between the poles and the equator is greatly modified by the rotation of the earth and by the nonuniform properties which are characteristic of its surface. Instead of a single circulation cell from equator to pole in each hemisphere, there are three latitudinal circulations, and there are also important longitudinal variations around each hemisphere.

The picture of the general circulation can be simplified somewhat by averaging the observed winds along each latitude, thus eliminating the longitudinal variations. Figure 5.2 is a schematic representation of the results. The horizontal flow at the earth's surface is shown in the center of the diagram; the net meridional circulation, at the surface and aloft, is depicted around the periphery. The component of the flow along meridians (north/south) has a speed, on the average, of less than a tenth

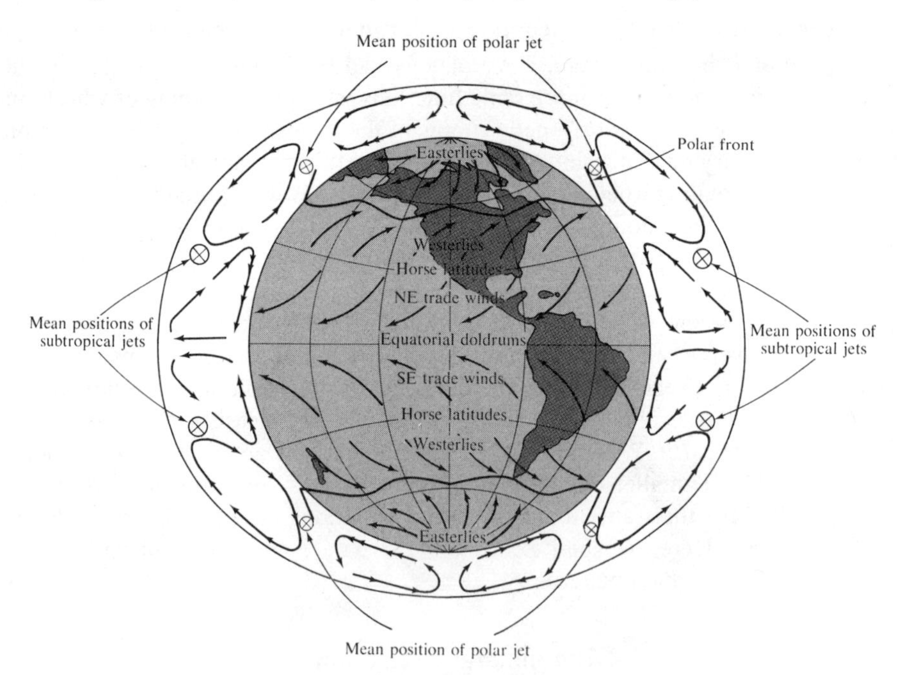

FIGURE 5.2 *Schematic representation of the general circulation of the atmosphere. Double-headed arrows in cross section indicate wind component from the east.*

of that along latitude circles (west/east), indicating the importance of the effect of the Coriolis force on this largest of scales.

Within the equatorial region are the *doldrums,* or intertropical convergence zone, a belt of weak horizontal pressure gradient and consequent light and variable winds. Here, also, is the region of maximum solar heating, and the surface air rises (as shown by the vertical circulation at the edge of the diagram) and flows both northward and southward toward the poles. This poleward flow at high levels is acted upon by the Coriolis force, turning the wind to the right in the Northern Hemisphere and to the left in the Southern Hemisphere. Thus, in both hemispheres, the poleward flowing air becomes a west wind and, at an average latitude of about 30 degrees, reaches a maximum speed which may exceed 100 mi/h. These are the *jet streams,* which are discussed in more detail later on.

At about 30 degrees north and south latitudes, some of the air descends again toward the surface. This is the region of the *horse latitudes* (so-called because Spanish sailing vessels carrying horses to the New World were occasionally becalmed in these areas of light winds and many of the animals had to be thrown into the sea because of the lack of food). Since the air is generally descending in these zones, there is little cloudiness or precipitation, and it is here that most of the world's great deserts are found.

Between the doldrums and the horse latitudes are wide belts where a portion of the previously equatorial air, having been cooled by its journey northward and dried out by its descent to the surface, returns again to the tropics. However, it does not flow directly southward, but is deflected by the Coriolis force, so that the wind moves from the northeast in the Northern Hemisphere and from the southeast in the Southern Hemisphere. These are the remarkably persistent *trade winds,* which obtained their name because of the important role they played in opening up the New World when ships were dependent upon sails.

From the horse latitudes, some of the descending air moves poleward at low levels, but the flow is deflected by the Coriolis force and the winds have a westerly component. These are the *prevailing westerlies.* During the days of sailing ships, they provided the motive power for vessels returning from North America to Europe. As the warm poleward-moving air reaches a latitude which varies from 40 degrees to 60 degrees, it encounters a cold flow from the pole. As a result of this encounter, a boundary is formed between the two masses of air known as the *polar front.** Here, the warm, light air from the horse latitudes is forced to rise over the cold, dense air from the pole, and a portion of the warm air returns at high levels toward the equator.

Poleward of the polar front are the *polar easterlies.* These winds bring the cold arctic and antarctic air from the polar regions toward the polar front, where they are warmed by their equatorward movement (and also in individual storms by mixing with the warmer air on the other side of the front). The air thus rises and returns toward the poles as a westerly flow aloft.

---

*We shall discuss the polar front in more detail later in this chapter.

From this description of the general circulation of the atmosphere, we see that there are two primary zones of rising air—in the tropics and in the region of the polar front. As might be expected, it is here that the principal areas of precipitation are found. Complementing these regions are the zones of descending air—in the horse latitudes and near the poles. Here, the precipitation is relatively light. We have already noted that the main deserts of the world are found in the horse latitudes, and while meteorological data near the poles are sparse, it is known that the precipitation in these areas is also small. However, because of the very low rate of evaporation of the ice in polar regions, it remains on the ground for long periods of time.

The preceding discussion represents an average or *mean* description of the atmospheric circulation for the entire year as a function of latitude only. The patterns of mean sea-level pressure during January and July, shown in Figure 5.3, give some idea of how the circulation varies over the surface of the earth during the year. Note that the subtropical high-pressure belt, associated with the accumulation of air in the horse latitudes, is not continuous around the hemisphere either in winter or summer, but rather is broken up into cells over the Atlantic and Pacific Oceans. These two cells are especially well defined in summer and are displaced several degrees of latitude farther northward in summer than in winter.

Since the winds blow clockwise around high pressure in the Northern Hemisphere, the *eastern* periphery of each cell is under the influence of relatively cool, dry northerly flow. Thus, the coastal areas of southwestern North America and Europe are favored by generally pleasant, rainless summers. In the interior of these regions (including northern Africa) are the major deserts. The *western* periphery of each high-pressure cell is associated with warm, moist flow from the tropics. Accordingly, locations such as the southeastern United States, as well as Hawaii, the Philippines, and southeastern Asia, are typically warm, with high humidity and frequent summer showers.

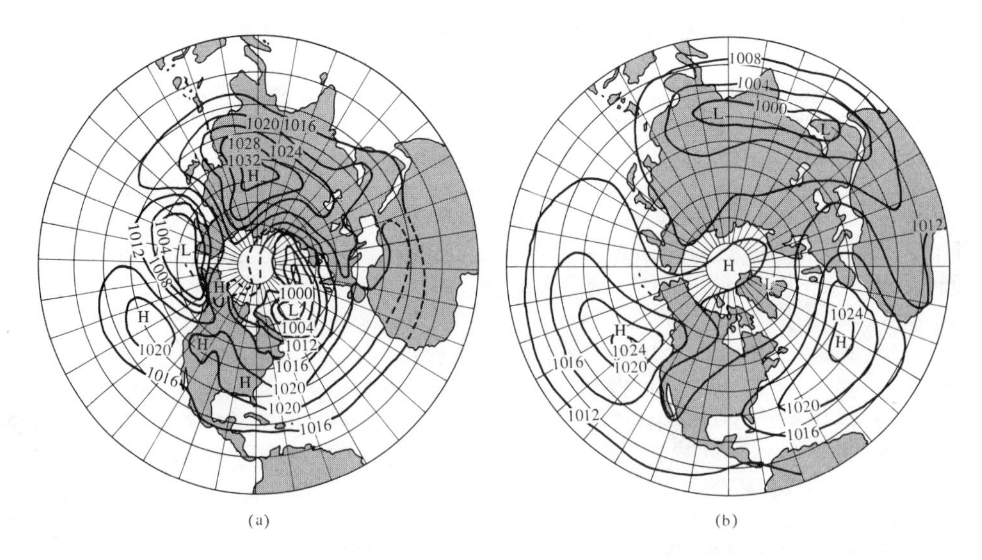

FIGURE 5.3 *Normal sea-level pressure in the Northern Hemisphere. (a) January; (b) July.*

Near the latitude of the polar front, where the relatively warm, moist prevailing westerlies meet the cold polar easterlies, are two low-pressure centers. These are well defined in winter, but they almost disappear in summer. Because of their location, they are termed the Aleutian low (in the Pacific) and the Icelandic low (in the Atlantic). They represent semipermanent "centers of action," where the major mid-latitude storms develop their greatest intensity.

In the interior of the North American and Asian continents, the low temperatures of winter result in increased density of the air at low levels and produce the cold high-pressure systems noted in those areas. However, in summer, temperatures are high, the air is less dense, and warm low-pressure systems prevail.

Figure 5.4 shows the corresponding flow at 500 millibars (about 18,000 feet) during the summer and winter in the Northern Hemisphere. At these levels, westerly winds dominate the region poleward of the horse latitudes, but they are more intense in winter than in summer, as evidenced by the closer spacing of the winter contours. The geographic center of the flow (lowest contour height) is not generally located at the pole, but it is displaced some distance away. In the Northern Hemisphere, the center of the lowest 500-millibars height is located over western Greenland, with a secondary center over eastern Siberia. These centers are the upper-level portions of the Icelandic and Aleutian lows, referred to earlier in this section.

Since the oceans in the Southern Hemisphere cover a significantly larger area than the oceans in the Northern Hemisphere, the temperature is much less affected by fluctuations due to continental influences. Thus, the pressure patterns, and therefore the large-scale winds, are much more symmetrical around the pole in the Southern Hemisphere than in the hemisphere north of the equator. Otherwise, the upper-level flow is similar, being consistently from the west throughout the region from about latitude 30° to near the pole.

Although the existence of jet streams had been postulated by theory much earlier, they were not actually observed until 1946, when high-flying military air-

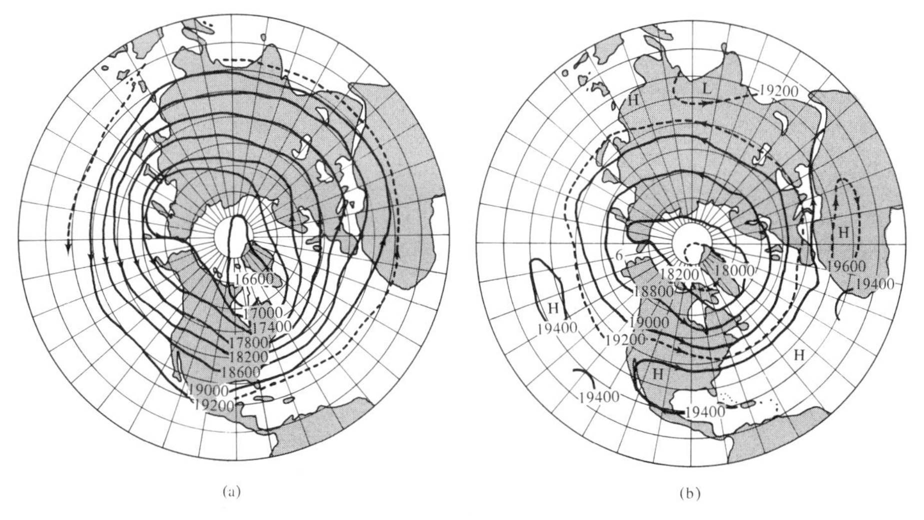

FIGURE 5.4 *Normal contours at 500 millibars in the Northern Hemisphere. (a) January; (b) July.*

craft encountered unexpected strong head winds against which they could make but little progress. Jet streams are narrow bands of high-velocity winds that meander, like great rivers, around each hemisphere at elevations extending from 4 to 5 kilometers to above the tropopause. The cores of these "rivers" of air are usually about 100 kilometers wide and 2 or 3 kilometers deep, and they flow at speeds as much as 100 knots faster than the air on either side of them. The location and intensity of jet streams change from day to day throughout the year, but they are associated with zones of strong horizontal temperature change and therefore follow the oscillations closely in position and strength of the polar front. In addition to the circumpolar jet stream, which is normally found between 35 degrees and 60 degrees latitude, a second, "subtropical jet stream" has been uncovered in the horse latitudes at very high elevations (9–13 kilometers). The subtropical jet stream does not meander over such a range of latitudes as does the circumpolar jet stream.

Basis for the general circulation Our knowledge of the general circulation is based on a sparse observation network. Very few observations have been made over the unpopulated areas of the earth—the oceans, mountains, jungles, Arctic, and Antarctic—which comprise most of the earth's surface. Nevertheless, sufficient information has been gathered to indicate that the general circulation is considerably different from that which might be expected for a uniform, nonrotating earth.

Clearly, a single thermal cell does not extend from equator to pole. The meridional (south–north) component that would be expected in a simple thermally driven circulation has become mostly "zonal" (along latitude circles) because of the influence of the Coriolis force. There is only one remnant of a clearly defined thermal circulation—that of the tropical cell between the equator and 30 degrees. There may also be a much weaker and less persistent *polar cell* over polar regions. But certainly between these two zones, and perhaps over the entire earth poleward of 30 degrees, there is no organized thermal circulation; rather, only frequent large eddies (cyclonic disturbances) that intermittently transport heat and momentum between the tropical cell and the polar regions. Note from Figure 5.2 that the mean flow within the middle-latitude cell shows meridional motion that is actually the opposite of what a thermal circulation should be like.

A study of the angular momentum of the atmosphere indicates that air moving poleward from the tropics would have an excessive westerly speed if there were not some mechanism by which its momentum were reduced. This can be seen by recourse to a basic principle of physics. The angular momentum of any body is given by $m\omega r$, where m is the mass of the body (in this case a parcel of air), ω is the angular velocity of the body, and r is the "spin radius" (i.e., the perpendicular distance of the body from the earth's axis).

Now if there are no torques (forces that cause turning, such as that applied to a pipe by a wrench), the angular momentum of the air parcel will not change. But if a parcel of air moves from a lower to a higher latitude at a constant distance from the earth's surface, its distance from the earth's axis (r) will decrease. Thus, if its angular momentum and its mass (m) remain constant, the angular velocity (ω) must increase. (This is like the skater who makes himself spin faster by pulling his arms in toward his body, thus concentrating his mass near the spin axis.) Since the earth's angular velocity is the same at all latitudes, if the parcel were originally spinning at the same rate as the earth's surface, it will be spinning at a faster rate than the underlying surface when it arrives at higher latitudes. This means that as an air parcel moves from the equator toward the pole, it will have acquired an additional speed from the west, relative to the earth. As an example, a parcel of air displaced from the equator to latitude 60 degrees would acquire a west-to-east speed of about 230 m/s (828 km/h)! Since such speeds are far greater than those ever observed, some mechanism must be responsible for slowing down the air which is transported from the equator to higher latitudes.

This mechanism is believed to be the cyclonic disturbances, or storms, of middle latitudes. The air in the easterly trade winds, slowed by the friction of the earth's surface, has its westerly angular momentum increased as it moves toward the equator. This acquired angular momentum, transported poleward by the tropical cell, is gradually absorbed by the great cyclonic storms of middle latitudes and, in turn, is dissipated at the earth's surface through friction. Thus, these storms embedded in the westerly flow of middle latitudes dissipate the excess momentum, much like the small turbulent eddies near the earth's surface diffuse a high concentration of smoke in the air.

While an overall analysis of the general circulation can thus be made from our current knowledge, a more detailed explanation of its time and space variations cannot yet be accomplished. This means that, since predictions of the weather for weeks or months in advance necessitate a better understanding of the behavior of this largest scale of motion, accurate day-to-day, long-range weather forecasting cannot be achieved at present. However, modern technological developments such as meteorological satellites, electronic computers, and advancements in meteorological theory all give promise that significant progress can be made toward this goal.

Cyclones and anticyclones

Embedded in the great circumpolar vortex of the general circulation are the cyclones and anticyclones of middle and high latitudes. These smaller-scale vortices tend to be masked by the averaging process used to analyze the general circulation, but an inspection of the flow for an individual day reveals their existence. For example, Figure 7.3(a) shows an intense cyclone centered off the southern New England Coast. As we shall see in Chapter 7, heavy snowfall produced by this storm brought a white Christmas to most of the Northeast. South of the cyclone, an anticyclone or high was centered over southern Louisiana, bringing clear and chilly weather to the South. Compared to the size of the circumpolar whirl (about 10,000 kilometers in diameter), cyclonic and anticyclonic eddies, such as those in Figure 7.3a, range from 1000 to 4000 kilometers in diameter.

The flow at 500 mb (about 5 km) at 0700 EST, Dec. 25, 1978, is shown in Figure 7.3(b). The solid lines in Figure 7.3(b) depict the height of the 500-mb pressure surface above the ground. Height contours on a constant pressure surface are analogous to isobars on a constant height surface, so winds tend to blow parallel to the height contours (see Appendix 3 for an explanation of the plotting model for the 500-mb chart).

A comparison of the surface [Figure 7.3(a)] and upper-air [Figure 7.3(b)] chart shows that the flow and pressure patterns aloft are somewhat simpler than those at the surface. The circulation is smoother and more wavelike aloft, with the basic flow being westerly over much of the United States. However, the simple circumpolar vortex that is found in the average flow at 500 mb [see Figure 5.4(a)] is not found on individual days. Rather, the simple pattern is perturbed by traveling waves that are intimately connected with surface cyclones and anticyclones. The troughs of the waves are associated with surface cyclones, while the ridges are associated with anticyclones. For example, the surface New England cyclone is associated with the upper-level trough that extends from Pennsylvania northwestward into central Canada and Alaska. We will see in a later section how these upper-level disturbances are the cause of surface cyclogenesis.

A comparison of Figures 7.3(a) and 7.3(b) illustrates the concept shown in Figure 4.4 and discussed on pages 60–63. The anticyclone over the Gulf Coast consists of cold air in the low levels. As discussed on page 62, such cold highs weaken with increasing elevation, and therefore no evidence of the high is found at 500 mb. In contrast, the northwest quadrant of the surface cyclone contains cold air; therefore the cyclonic circulation increases with elevation in this region. Finally, we note that cold air aloft is associated with low heights, while warm air aloft is accompanied by high heights [Figure 7.3(b)].

The large cyclonic and anticyclonic eddies that move in a general easterly direction around each hemisphere dominate the flow over much of the earth between about 30 degrees and 75 degrees latitude. As was mentioned earlier, they are more significant transporters of heat and momentum between low and high latitudes than is the mean meridional (south-north) flow of the general circulation. This efficient transport occurs because southerly winds around the eastern semicircle of cyclones carry warm air northward, while on the western semicircle, northerly winds transport cold air southward. The opposite circulation occurs around anticyclones, so that both pressure systems act as efficient heat exchangers between low and high latitudes.

The cyclonic whirls are the "storms" of middle latitudes. In the temperate latitudes they produce much of the winter precipitation. Around their low-pressure centers, the air circulates in a counterclockwise direction in the Northern Hemisphere and in a clockwise direction in the Southern Hemisphere. The masses of air that circulate around them are generally heterogeneous with respect to temperature and moisture, having come from different geographical areas; as a result, there exist sharp transition zones separating warm, moist air from cold, dry masses. These storm systems go through a complex life cycle which will be discussed in more detail later in this chapter.

Winds in the anticyclone circulations blow clockwise around their high-pressure centers in the Northern Hemisphere and counterclockwise in the Southern Hemisphere. Within these whirls, the air is slowly subsiding at the rate of 10–15 cm/s and "fair weather" generally prevails. The air masses of which they are composed are generally homogeneous with respect to temperature and moisture.

The monsoon

A large-scale example of a thermal circulation is the *monsoon*. Derived from the Arabic word for season, it refers to a wind circulation that is seasonal in character. During the winter, when continents are colder than the oceans, air flows outward from the continents; while during the summer, when the continents are warmer than the oceans, the flow is inward.

Seasonal precipitation amounts are closely linked to the monsoon at places where the circulation is well developed, as in the Asian continent. The summer monsoon brings in moist oceanic air to the continent, where it rises, leading to condensation. But in the winter monsoon, precipitation is much less likely because over the continents the air is subsiding and streaming outward to the oceans.

The most intense monsoons are those produced by the large Asian land mass. The climate of southern Asia, protected from the north by the towering Himalayas, is largely determined by the monsoon. In summer, the southerly winds over the northern Indian Ocean deposit the heaviest rainfall in the world along the southern Himalayan slopes. In winter, the prevailing northeast winds are dry and there is little rain. The onset and duration of the summer monsoon are of great significance to agriculture in southern Asia. The geographic distribution, intensity, and duration vary considerably from year to year. Recently, meteorologists have launched an intensive investigation of the causes for these fluctuations, although they still cannot be forecast with any degree of certainty.

North America also experiences a monsoon circulation, although it is not nearly as strong as the Asian one and tends to be obscured by migratory cyclones and fronts. Its most noticeable effect is that, in summer, the hot interior draws in moist tropical air from the Gulf of Mexico and Caribbean Sea. The summer thunderstorms over the arid highlands of the southwest can be attributed partially to the influx of moist air from the Gulf of Mexico because of the monsoon circulation.

Land and sea breezes

A coastline is a sharp boundary between surfaces having greatly different temperature variations. The sea, because it is constantly being stirred, has a relatively small diurnal temperature change compared to the adjacent land surface. As a result, a large temperature difference can develop across the coastline during certain times of the day. In the tropics throughout the year, and at higher latitudes during the summer, the land-sea temperature gradient on a fairly calm, clear afternoon can reach 15°C over a distance of less than 50 kilometers. At night, the temperature difference may be reversed (ocean warmer than land), although normally not nearly so pronounced.

These land-sea temperature differences lead to the creation of a thermal circulation such as that shown in Figure 5.5. The daytime landward flow is know as a *sea breeze,* while the seaward flow at night is called a *land breeze*. Unlike the monsoon, the sea breeze only occasionally causes rainfall, although it frequently induces cloudiness if the marine air is forced upward over inland slopes. In the

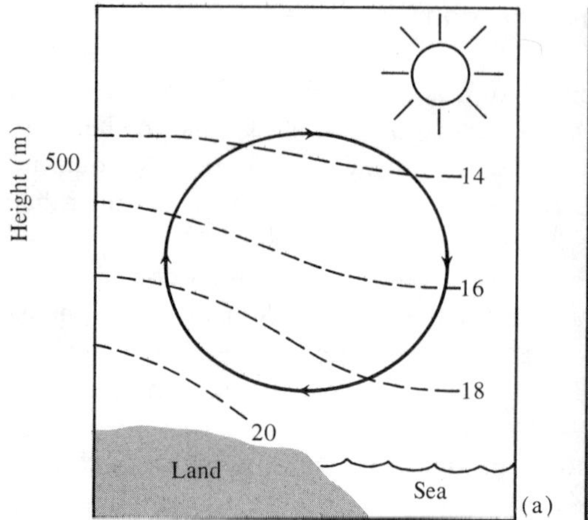

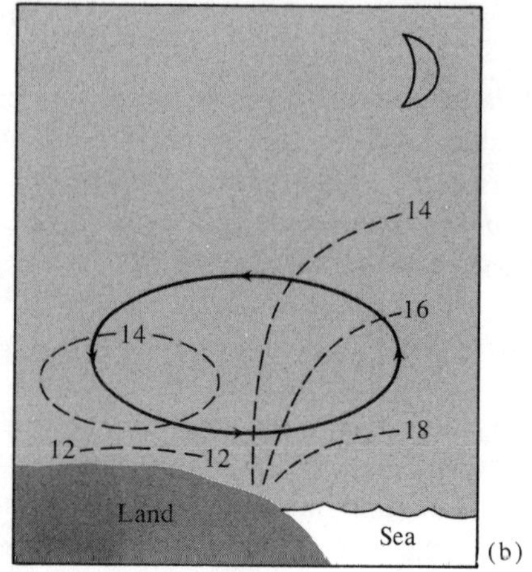

FIGURE 5.5 *Sea-breeze (a) and land breeze (b) circulations. Dashed lines are isotherms in degrees Celsius. Solid lines indicate direction of flow.*

tropics it may occur all year round, but at higher latitudes it is mostly a summer phenomenon.

The sea breeze usually begins to develop three or four hours after sunrise, because the air over the land heats up more rapidly than air over water. By 1 or 2 P.M., when it has reached its peak intensity, the circulation cell usually extends both inland and seaward about 20 kilometers, although it has been found to penetrate inland as much as 60 or 70 kilometers. The entire circulation cell, including the upper, seaward flow, is not normally more than 1 kilometer deep, although in the tropics it may reach to 3 kilometers or more. The surface wind is usually gusty and constantly shifting in direction. As the forward edge of the sea breeze passes over a point in its landward penetration, the relative humidity increases and the temperature decreases sharply (the former by 40 percent or more and the latter 5°C or more in less than an hour). Sometimes fog or low stratus clouds may accompany the sea breeze; there are places where the coastal water is extremely cold, such as along the Peruvian coast, where the forward edge of the sea breeze is so sharply defined that the fog appears as a solid wall.

As the land cools in the evening, the sea breeze dies, and between about 7 and 10 P.M., there is little evidence of it. The land breeze, much weaker than its daytime counterpart, will normally begin at 10 or 11 P.M. and reach its maximum development near sunrise. The principal effect of the land breeze is to prevent the air over the land from cooling quite as much as it otherwise might.

The sea breeze plays an important role in moderating the temperature of narrow strips of land along seacoasts and lake fronts. The breeze created by lakes is generally much less intense and has a smaller width and depth. Along the shores of the Great Lakes, for example, the inland penetration is usually not more than a few kilometers. However, it offers a welcome relief from the summer heat for residents who live close to the shore.

Mountain and valley winds

Along mountain slopes, a thermal circulation occurs that also has a diurnal cycle. During the daytime, from about three hours after sunrise until sunset, an upslope wind, called a valley wind, blows. Between about midnight and sunrise, an opposite, downslope wind, called a mountain wind, occurs. The mountain-valley winds are most pronounced on clear, summer days, when the prevailing winds are weak.

The mountain-valley circulation is produced because the air in contact with the slope is either warmer and less dense (daytime) or cooler and denser (nighttime) than air at the same elevation over the valley. As a result, the air over the slopes rises during the day and sinks at night. Of course, the intensity of the flow and the specific direction at any point depends on the degree of slope and the configuration of the valley. Mountain and valley winds are best developed in wide, deep valleys.

The rising air currents along mountain slopes are a familiar phenomenon to every mountain climber. Frequently, cumulus clouds and showers form over summits in the ascending, expanding air. The depth of these rising currents above the slopes is usually between 100 and 200 meters.

Katabatic winds

All downslope, drainage-type winds are referred to as *katabatic* winds. Most are weak, usually not exceeding 10 km/h, and are significant primarily because they cause cold air to drain into the valley, producing lower night temperatures in the valley than on the mountainside.

There are some very strong drainage (katabatic) winds, but most of these are set into motion by the large-scale or prevailing flow. One of the strongest of these, the glacier wind, may attain destructive violence. It occurs when air is cooled as it moves across snow fields on high plateaus; at the edge of the plateau the air cascades downward. At some places, such as along the fjorded coasts of Norway, Greenland, and Alaska, deep canyons channel the flow, thus augmenting the speed considerably. These winds blow during both the day and night. Another example is the *bora* wind, which sporadically brings in cold air down rather steep slopes to the usually warm Adriatic Sea. The bora is in intermittent wind, gusts of 50 to 100 km/h being interspersed with calms. Where it reaches the sea, it produces great waves and kicks up spray in great quantities. A similar cold wind, the *mistral,* occurs along the French Mediterranean coast.

Föhn winds

The Föhn wind is a downslope wind that occurs in many mountainous areas, but it is not caused by drainage of dense air. It is a warm, very dry, erratic wind that sometimes appears along the lee slopes of a mountain ridge. It occurs when the prevailing winds in warm, moist air are directed against a mountain. The forced ascent causes thick clouds to form and, on occasion, heavy orographic precipitation. During most of the ascent, cooling proceeds at the *moist* adiabatic rate (4° to 5°C per kilometer) and by the time the air reaches the peak level, much of its moisture has been removed.

After crossing the ridge line, some of the air descends along the lee slopes, warming at the *dry* adiabatic rate (10°C per kilometer). When it arrives near the bottom of the mountain, the air is very warm and dry, having been heated by the latent heat of condensation.

Although föhn winds are observed along many mountain ranges in the world, some of the most extreme cases occur along the eastern slopes of the Rocky Mountains. Here the phenomenon usually goes by the name of *chinook,* the Indian territory from which they seemed to come. The Indians commonly referred to it as the "snow eater" because its extreme dryness and warmth could melt and evaporate as much as a meter of snow in a day. The chinook wind frequently forces out the cold air that lies along the eastern slopes and the temperature has been observed to rise by 50°F to 60°F in half a day after its arrival.

5.3 Thunderstorms

Thunderstorms

As we have already pointed out, vertical motion in the atmosphere is the key to many of the characteristics of weather. Upward motion results in expansion, cooling, and eventual condensation of the water vapor in a stream of air; the release of latent heat is often an important factor in accelerating the convection by increasing the buoyancy (instability) of the air. Downward motion results in compression, warming, and therefore an increase in the air's capacity for water vapor. We have seen, also, that convective patterns come in a large variety of sizes, the smaller ones nesting within the larger ones. In general, the maximum vertical velocity observed is inversely proportional to the size of the circulation: the large patterns have relatively feeble vertical motion while many of the small circulations have vertical motion equal to that in the horizontal.

Cloud types are closely related to the strength of the vertical motion. Stratified clouds, which sometimes extend unbroken over thousands of square miles, occur in gently ascending air (almost always less than 20 cm/s). Cumuliform clouds, on the other hand, occur as isolated cloud masses (individual elements rarely cover more than 75 km^2) and contain within them upward motions as strong as 100 km/h. Stratified clouds form in air in which the buoyancy forces are weak or even oppose vertical motion above the thin layer in which the clouds form. For example, a wind blowing up a mountain slope may lead to condensation, but a temperature inversion may prevent vertical development of the clouds. Cumuliform clouds—those with great vertical development—are associated with instability.

The rapidity with which convective cells can develop in an unstable atmosphere is illustrated in the photographs of Figure 4.18. Of the many individual cumulus convective cells appearing on the horizon in the first photograph of Figure 4.18, one mushroomed vertically into a mature storm in only 18 minutes. It is not usual that isolated convective "cells" of this sort can be identified; normally, there is a tendency for adjacent cells to develop and join together. Frequently, there are great masses or lines of thunderstorms extending over 80 kilometers or more, but a single "cell" has a diameter of about 8 kilometers.

Studies have shown that there are three characteristic stages in the life cycle of a typical thunderstorm cell. These are illustrated in Figure 5.6. The initial, *cumulus* stage usually lasts for about 15 minutes. During this period, the cell grows laterally from 2 or 4 kilometers in diameter to 10 or 15 kilometers, and vertically to 8 or 10 kilometers. Note from Figure 5.7(a) that the updraft is strongest (about 30 km/h) near the top of the cloud. Air is entering the cloud through the sides at all levels. The upward motion is actually greater than the horizontal speed, which is the reverse of what is found in larger-scale atmospheric circulations.

The *mature* stage (Figure 5.6) begins when rain falls out of the cloud base and usually lasts for 15–30 minutes. During this stage, the size of drops and ice crystals in the clouds grows so large that the updrafts can no longer support them, and they begin to fall as large drops or hail. The frictional drag of the precipitation gradually slows the updraft and, in one part of the cell, a strong downward motion develops because of the precipitation and "entrainment" of cooler air from outside the cloud. Near the center and top of the cloud, upward motion is still strong, however; speeds as high as 100 km/h have been observed. Note also the strong outflow below the base of the cloud. When this downdraft meets the ground it spreads away from the thunderstorm. It is for this reason that gusty, cool winds usually precede the actual arrival of a thunderstorm.

The mature stage is the most intense period of the thunderstorm. Lightning is most frequent during this period, turbulence is most severe, and hail, if present, is

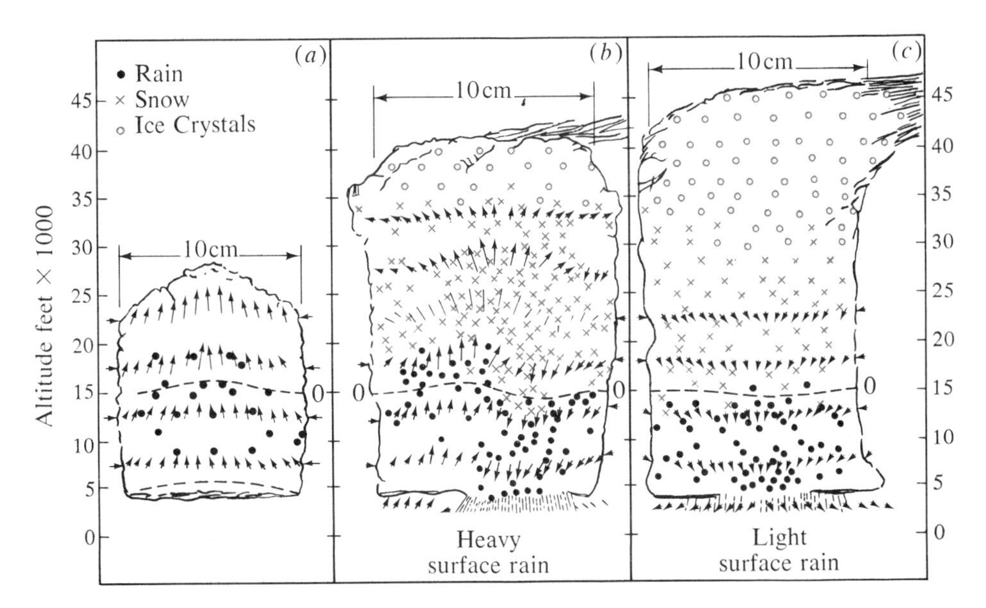

FIGURE 5.6 *Life cycle of a typical cumulonimbus cell. Schematic 2-dimensional representation shows the relative wind velocities, temperature, and distribution of liquid and solid water.*

most often found in this stage. The cloud reaches its greatest vertical development near the end of this stage, usually reaching about 10 kilometers and sometimes penetrating the tropopause to altitudes greater than 15 kilometers.

The final or *dissipating* stage begins when the downdraft has spread over the entire cell. With the updraft cut off, the rate of precipitation diminishes and so the downdrafts are also gradually subdued. Finally, the last flashes of lightning fade and the cloud begins to dissolve, perhaps persisting for a while in a stratified form.

Severe thunderstorms, which often spawn tornadoes, are considerably more complex than the more or less isolated thunderstorms of moderate intensity described previously. When the atmosphere is very unstable and abundant moisture exists in the low levels, thunderstorms may organize themselves into mesoscale circulations with typical diameters of 20 to 40 kilometers. As the thunderstorm cells pump low-level air into the upper troposphere at velocities of 40 to 80 km/h, low-level air flows in to compensate for the vertical motion. Through the conservation of angular momentum, the entire thunderstorm system may begin to rotate, producing a mesoscale cyclone. Such a circulation is shown in Figure 5.7, which shows the winds at a height of 0.3 km in a thunderstorm system in Oklahoma on June 8, 1974. The winds in Figure 5.7 were diagnosed by dual-Doppler radar measurements. Doppler radars emit a radio signal which is reflected by precipita-

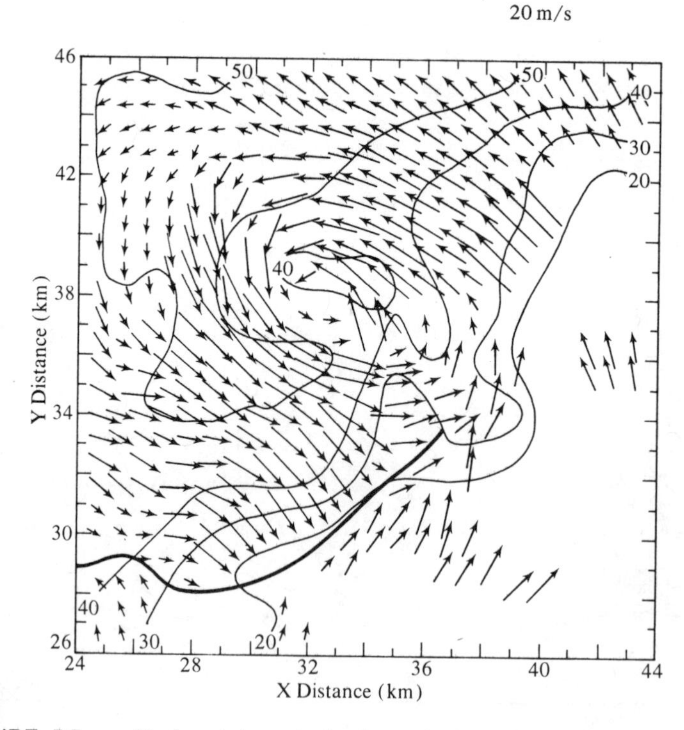

FIGURE 5.7 *Horizontal perturbation wind analysis of severe thunderstorm near Harrah, Oklahoma. Mean flow has been subtracted. Contour lines denote relative radar reflectivity. (E. Brandes,* Jour. Appl. Meteor. *April 1977.)*

tion particles back to a receiver. The speed of the particle toward or away from the source of radiation causes a change in the frequency of the radiation (the Doppler shift). Although one radar can give only the velocity component along a line from the emitter to the particle, the 3-dimensional flow can be reconstructed if there are two Doppler radars separated by some distance and observing the same particles.

The winds in Figure 5.7 show strong rotation, with winds exceeding 40 m/s (144 km/h) in places. The wind shift from southeast to northwest in the southern part of the storm (indicated by a heavy black line) is the gust front and marks the leading edge of cool air originating from a downdraft.

From Doppler radar analyses such as the one presented in Figure 5.7, a reasonably complete 3-dimensional picture of the circulation associated with long-lived severe thunderstorms is emerging (Figure 5.8). In Figure 5.8 a storm is moving toward the east and is being continually supplied with warm, moist, low-level air around its leading edge. In the updraft fed by this inflow, condensation produces rain below the freezing level and ice at higher levels. To the rear of the storm, some dry middle-level air is incorporated into the storm. As evaporation of rain cools this air, it becomes negatively buoyant and sinks. When the resulting downdraft reaches the ground, it spreads out and forms the gust front.

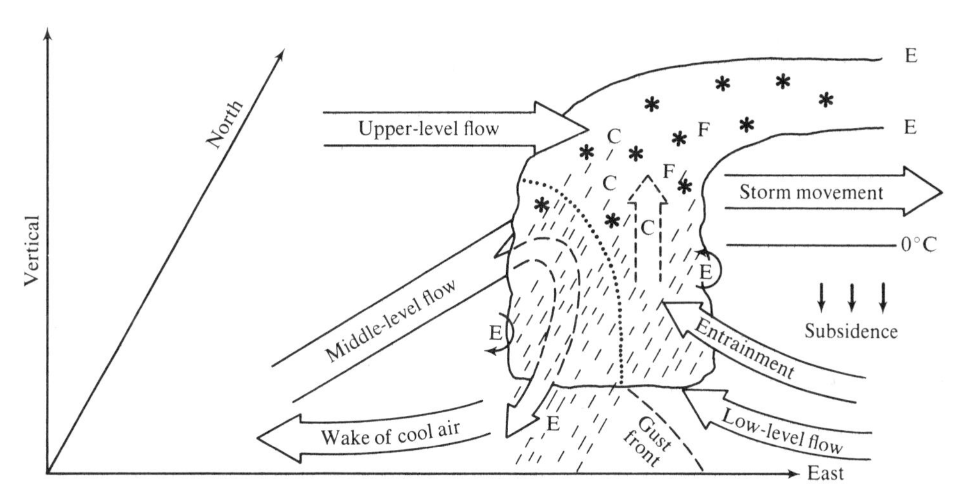

FIGURE 5.8 *Schematic diagram showing types of precipitation and flow in severe thunderstorms.*

Thunderstorms generally occur within moist, warm (maritime tropical) air masses that have become unstable either through surface heating or forced ascent over mountains or fronts. In the United States, practically the only source region of this air mass is the Gulf of Mexico and Caribbean. Note how the geographic pattern of thunderstorm incidence shown in Figure 5.9 is correlated with both the distance from the source region and topography.

A *thunderstorm* is, as the name implies, a storm accompanied by thunder and, therefore, lightning. It occurs in the cumulonimbus cloud. As Benjamin Franklin demonstrated in 1750, lightning discharges are giant electrical sparks. Cumulonim-

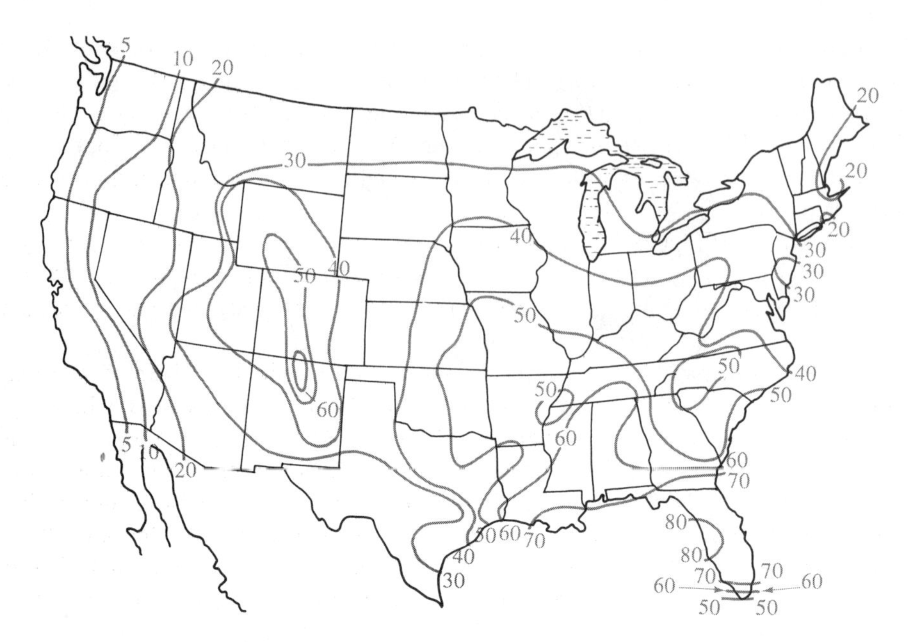

FIGURE 5.9 *Average annual number of days with thunderstorms.*

bus clouds, therefore, are great natural electrical generators. Like man-made machines, such as batteries, the cloud produces "poles" of concentrations of positive and negative electricity.

An important question that is still not adequately answered is how the powerful convective currents in these clouds produce the electrical charge and then separate the positive from the negative electricity. The lower part of a thundercloud has a concentration of negative charge, while the upper part is largely positive. The process that produces and separates charges must involve the water and ice particles in clouds. Some suggested processes follow. (1) Friction between the ice particles formed near the top of such clouds would cause the ice to become negatively charged. Large ice crystals that fall would then carry negative electricity downward, leaving the upper portion positive. (2) Water droplets, when they form, tend to attract negative ions. (Ions are molecules that have become charged through the loss or gain of an electron. In the atmosphere, ions are produced by the radiations from radioactive material in the soil, by cosmic rays from the sun, and by combustion, friction, and splitting of water drops in sprays.) (3) If a cloud has already a predominance of positive charge near the top and negative near the bottom, so that an electrical potential exists, then any drop will tend to distribute its internal charge. Thus, the bottom portion of the drop is positive and the upper portion is negative. If a current of air is sweeping upward past the drop, negative ions will be captured by the bottoms of the drops (which face the air current) more readily than will the positive ions. The rising air currents will therefore arrive near the cloud top with their negative ions depleted, or, in other words, with a positive charge.

Regardless of how the thundercloud does it, the fact remains that enormous potential differences are generated within clouds and between clouds and ground. Just before a discharge, the electrical potential gradient is of the order of 3000 volts per centimeter and potential differences between the extremities of flashes reach hundreds of millions of volts. A typical thunderstorm dissipates electrical energy at an average rate of about a million kilowatts.

Special photographic techniques have shown that individual lightning discharges actually consist of multiple strokes, each lasting about 0.0002 second, with about 0.0001 second between successive strokes. The air along the lightning channel is heated momentarily to about 15,000°C (compared to the sun's surface temperature of about 6000°C); this causes a very rapid expansion of air, which in turn results in the deep sound called *thunder*. The rumbling of thunder occurs because sound is generated over a long discharge path, so that sound waves travel over many different paths to the observer, and much of the sound is reflected. The approximate distance from a thunderstorm can be computed by noting the time elapsed between a flash of lightning and the arrival of the sound wave by using the average speed of sound (330 m/s, or 20 km/min).

The old proverb that lightning does not strike twice in the same place is, of course, untrue. Tall towers and buildings are repeatedly struck by lightning; Franklin's lightning rod protects such structures by providing a low resistance conductor of the electrical current to the ground. One should always avoid being near an isolated, high target during a thunderstorm. Do not get caught under a tree or on an open golf course during a storm.

5.4 Vortices

The more violent aspects of weather are associated with cyclonic rotating whirlpools of air called *vortices*. There are three principal types of *vortices* in the atmosphere; the *wave cyclone* of middle and high latitudes is the largest weather-producing vortex, but it is usually not the most violent. Typically, it has a diameter of about 2000 kilometers, and the wind near the earth's surface usually does not exceed 70 km/h (45 mi/h). Smaller in size (average diameter about 700 kilometers) and much more destructive is the *tropical cyclone*. The maximum surface wind speed is sometimes more than 200 km/h (125 mi/h). Fortunately, the tropical cyclone spends most of its life on the oceans where it does little harm. The smallest vortex, but the one with the most powerful punch, is the *tornado*. The intense rotation of this vortex is confined normally to a diameter of a kilometer or less, but its wind speed can reach 300 km/h (200 mi/h).

The extratropical wave cyclone is the best understood of these three vortices. By far the most common storm, it is found principally in the temperate latitudes and is large enough so that many of its characteristics can be ascertained by upper-air meteorological observations with an average spacing of 400 km. Less is known about the formation and structure of the more violent tropical cyclones and tornadoes. In the case of tropical cyclones, which form over the oceans and spend most of their lifetimes there, reports are too scanty to properly pin down their dynamics. Although tornadoes occur principally over continents, they are so small and have such a short lifetime (rarely longer than one hour) that they usually fall between the observation points of the standard land network of weather stations.

Low pressure lies at the center of all three of these strongly rotating vortices. When air converges toward a point, as it does in the case of a low-pressure center, its speed of rotation increases through the conservation of angular momentum. This increase in speed requires a greater centripetal force to make the air turn in even tighter circles. This centripetal force is the horizontal pressure gradient force, directed toward the center of the vortex (Figure 5.10). Because extratropical and tropical cyclones are large enough to be affected by the earth's rotation, their winds are always cyclonic* about the center of low pressure. Small tornadoes (and their wet cousins, waterspouts) are too small to be affected by Coriolis forces and anticyclonic winds have been observed in these storms.

Because the centripetal force is produced by the pressure gradient force for circulations about lows (Figure 5.10), and there is no theoretical limit to the magnitude of the pressure gradient force, the speed of winds around lows is nearly unlimited. Around highs, however, the balance of forces is such that the centripetal force must be produced by an excess of the Coriolis force over the pressure gradient force. Thus balanced flow around highs can only occur on large scales. Furthermore, since the required centripetal force for a given radius increases as the square of the wind speed, while the Coriolis force only increases as the first power of the wind speed, balanced flow around highs can only occur at relatively low wind speeds.

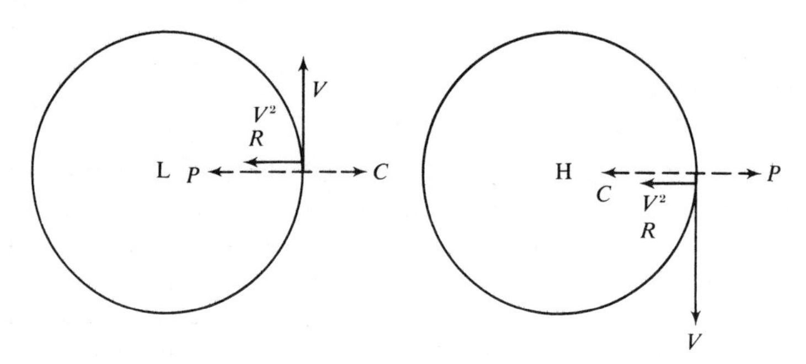

FIGURE 5.10 *Balance of forces around cyclones and anticyclones. P is pressure gradient force, C is Coriolis force, and* V^2/R *is required centripetal force.*

5.5 Air Masses, Fronts, and Wave Cyclones

In the discussion of the general circulation, mention was made of the importance of the large, essentially horizontal eddies in producing an exchange of air between the polar regions and the tropics. These large whirls are most active in the middle latitudes, where they are the chief weather producers. Periodically (in some places, every few days in winter), these waves cyclones, as they are called by meteorologists,

---

*A cyclonic circulation is one in which the rotation is the same as that of the local horizon about the vertical. Thus cyclones rotate counterclockwise over the northern hemisphere and clockwise over the southern hemisphere.

develop along the boundary between warm and cool streams of air, later sucking these streams of air toward their centers; in the process they transport "cold" equatorward and "heat" poleward. The prognosis of the formation and development of these cyclones is one of the principal tasks of the short-range forecaster in the polar and temperate regions of the earth. Two concepts are important in the explanation of these giant storms: *air masses* and *fronts.*

Air masses

An *air mass* is a huge body of air, extending over thousands of kilometers, within which the temperature and humidity change gradually in the horizontal; i.e., there are no sharp horizontal changes of temperature or humidity. Air masses are created principally within the anticyclonic flow of the subtropical and polar high-pressure belts. The air circulates slowly in these systems over surfaces of fairly uniform properties and gradually acquires thermal and moisture characteristics representative of these surfaces. For example, the air flowing around the semi-permanent Atlantic anticyclone very quickly acquires the warmth and moisture of such water bodies as the Caribbean and Gulf of Mexico. Cold air masses, such as those that form over the frozen surfaces of northern Canada in winter, take somewhat longer to form, but under fairly stagnant conditions horizontal homogeneity can exist to a 3-kilometer or 4-kilometer depth.

Air masses are classified according to their source region — *polar* or *tropical, maritime* or *continental.* The chief air masses that affect the weather of North America are continental polar(*cP*), maritime polar (*mP*), and maritime tropical (*mT*). The origin of continental polar air masses is northern Canada. In winter, the *cP* air mass is dry and stable before it moves out, but when it moves southward over the United States it is heated from below and its stability decreases. The portion that traverses the Great Lakes picks up moisture, which frequently results in snow showers along the eastern shores of the lakes and in the Appalachian Mountains. Occasionally, this *cP* air may penetrate the Rocky Mountain range.

The maritime tropical air that affects the United States generally comes from the Gulf of Mexico. In winter, maritime polar air sweeping out of the Pacific is largely responsible for the winter rains of the west coast of the United States. As it strikes the coastal range and then the Rockies, the forced lifting causes heavy rain and snow over these barriers.

After it has left its source region, an air mass can be further characterized by its temperature relative to the surface over which it is traveling. An air mass is said to be *cold* if it is colder than the underlying surface and *warm* if it is warmer than the surface. A cold air mass will be heated from below, thus the lapse rate will increase, while a warm air mass will lose heat to the underlying surface and its lapse rate will decrease (it will become more stable).

Fronts

Across a boundary separating air masses of differing properties there may exist a sharp contrast of temperature and humidity. Such a boundary, where air masses "clash," is called a *frontal zone* or, more commonly, a *front*. The name "front" was coined by the Norwegian meteorologists who first developed the polar front theory during World War I, possibly because the oscillations of the boundary, with periodic flareups of weather along it, reminded them of the long battle line in Europe with its intermittent activity.

The front separating two air masses slopes upward over the cold, denser air. This is illustrated in Figure 5.11, a typical vertical cross section of fronts over the center of the North American continent. Note that the vertical scale is greatly exaggerated. The average slope of fronts is only about 1:150, ranging from as little as 1:250 to as steep as 1:50. The width of the front — the transition zone between air masses — is usually about 50–100 kilometers, but on the scale of distances that we are considering, such a width is closely approximated by the thickness of a heavy line drawn on a weather map.

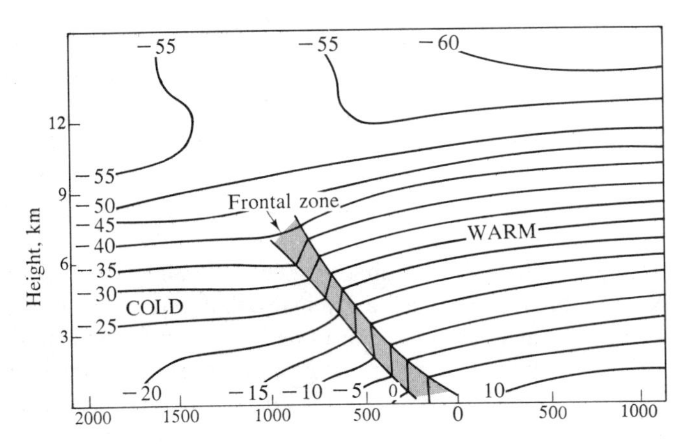

FIGURE 5.11 *Vertical cross section of a front. (North–south cut, North America, on a winter day; isotherms in °C.)*

The boundary between the warm and cold air masses must always slope upward over the cold air. This is because the cold air is denser fluid. (Imagine two fluids such as water and oil, side by side, separated by a partition. If the partition is removed, the heavier water will slide beneath the oil.) Now if either the warm air is moving against the wedge of cold air or the wedge is pushing under the warm air, there will be forced lifting. In either case, cooling due to expansion may lead to condensation and possibly precipitation over the frontal surface.

Wave cyclones

The surface weather pattern associated with migratory cyclonic depressions of the middle latitudes has been known for about 80 years. The polar front model associated the formation and maturation of these storms with undulations of the frontal boundary. A front separates air masses of different densities. The air masses flowing side by side may develop zones of strong wind "shear" between them; i.e., the currents of air on both sides of the boundary may have different velocities. When the flow aloft is undisturbed, fronts show little development even though temperature contrasts may be large across the front. Development of surface cyclones along fronts occurs when an upper-level disturbance approaches a front. The upper-level patterns of convergence and divergence produce surface pressure rises and falls, respectively, and these surface pressure changes generate low-level circulations (Figure 5.12). As the low-level cyclone intensifies, the surface front is progressively deformed as cold air is swept southeastward behind the low and warm air moves northeastward. Successive configurations of the front resemble a developing and then breaking wave, hence the name "wave cyclone" is frequently used to describe these systems.

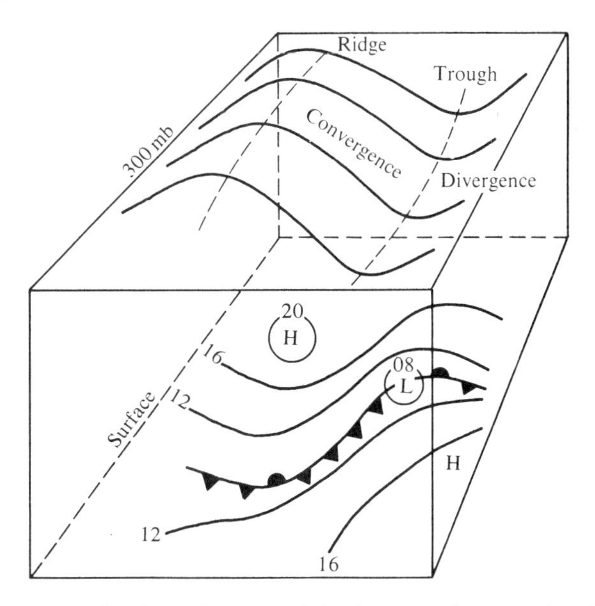

FIGURE 5.12 *Surface front and isobars and upper-level flow for developing cyclone. Upper-level divergence over the low causes pressure falls; convergence over high produces pressure rises.*

As the wave develops, low pressure forms at its apex (Figure 5.13) and both the warm and cold currents move in a cyclonic pattern around it. To the left (in the figure) of the apex, the front is advancing toward the warm air, and this segment of the front is called the *cold front;* to the right of the apex, the front is receding from the warm air and so this segment is called the *warm front.* The warm air between the fronts is known as the *warm sector.*

Figure 5.13 represents an idealized wave cyclone, the view looking downward on the earth shown in the center and vertical cross sections taken a little south (bottom drawing) and a little north (top) of the apex. Imagine the entire system moving toward the right (eastward), as is normally the case. If you were standing to the east and south of the apex, ahead of the warm front, the first sign of the approaching system would be high cirrus clouds. As time goes on, the wisps of cirrus thicken to cirrostratus clouds; these often cause halos (rings around the sun

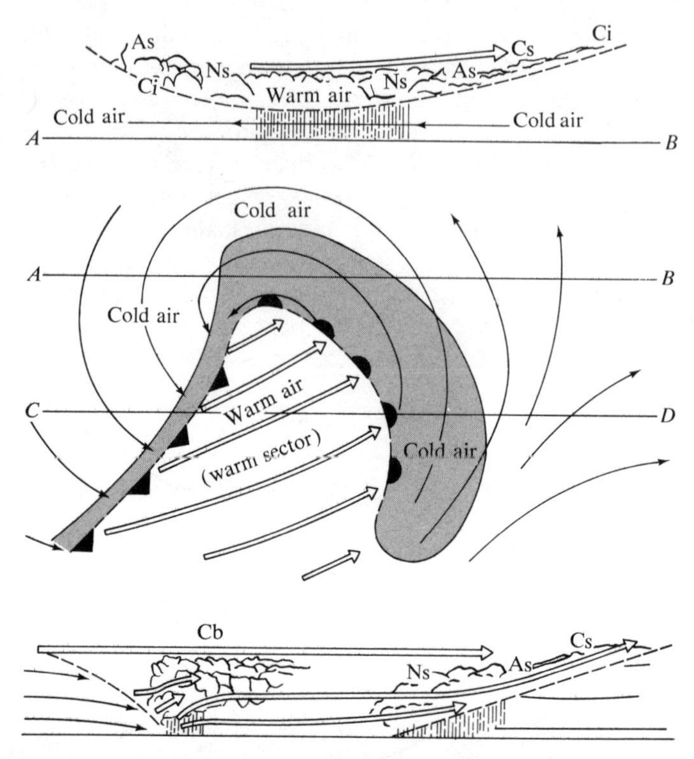

FIGURE 5.13 *The wave cyclone model. (After J. Bjerknes and H. Solberg.) Center drawing, horizontal plane view; top, vertical cross-sectional view just north of wave apex (line AB); bottom, vertical cross-sectional view across warm sector (line CD). (For abbreviations of cloud-type names, see page 29; arrows depict air flow.)*

or moon), a sure sign of rain within 24 hours, according to a well-known proverb. Gradually the clouds lower and thicken to altostratus. The pressure falls, and the wind increases and backs (changes direction in a counterclockwise direction), as the low center gets closer. The temperature begins to rise slowly as the frontal transition zone approaches. Within 300 kilometers of the surface position of the front, precipitation begins, either in the form of rain or snow. After the warm front passes, the precipitation stops, the wind veers into the southwest (changes direction in a clockwise direction) and the pressure stops falling. Within the warm sector, the weather depends largely on the stability of the warm air mass and the surface over which it is moving; there may be showers or almost clear skies.

The type of weather accompanying the passage of the cold front depends on the sharpness of the front, its speed, and the stability of the air being forced aloft. Usually there are towering cumulus and showers along the forward edge of the front. Sometimes, especially in the midwest during the spring, severe *squalls* precede the front. But in other cases, nimbostratus and rain extend over a zone of 75–100 kilometers. After the frontal passage, the wind veers sharply and the pressure begins to rise. Within a short distance behind the cold front, the weather clears, the temperature begins to fall, and the visibility greatly improves.

The early genesis stages of the wave cyclone normally take between 12 and 24 hours. Subsequent development of the wave, shown in Figure 5.14, takes an additional two or three days. As the wave breaks, the cold front begins to overtake the warm front. This process is called *occlusion* and the resulting boundary is called an *occluded front*. The vertical cross sections of Figure 5.15 illustrate that either the cold front can move up over the warm front (warm-front type occlusion), or it can force itself under the warm front (cold-front type). The occluded front is the boundary that separates the two cold air masses.

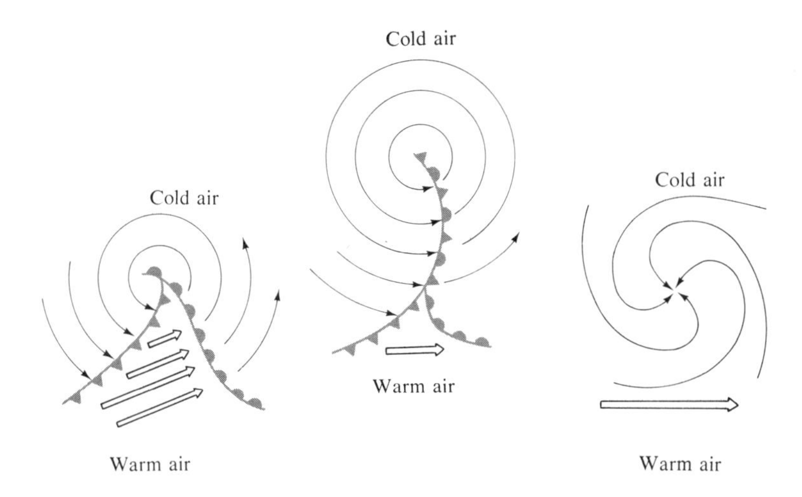

FIGURE 5.14 *Later stages in the development of a wave cyclone. (Northern Hemisphere.)*

The maximum intensity of the wave cyclone, in terms of horizontal pressure gradient and wind velocity, normally occurs during the occlusion process. This happens because occlusion produces a redistribution of the air masses. The denser,

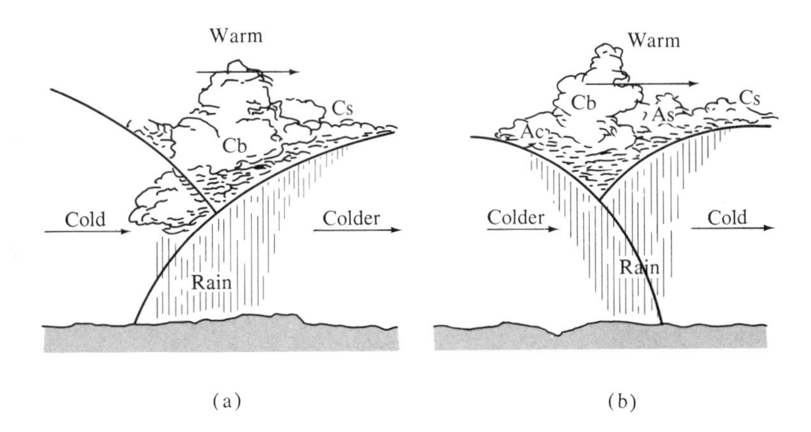

FIGURE 5.15 *Vertical cross section of occlusions. (a) Warm-front type; and (b) Cold-front type. (Arrows indicate displacement of air masses.)*

cold air moves in at the surface of the system and less dense, warm air is forced aloft. This change in the distribution of mass within the eddy results in a loss of potential energy (more light air aloft, and more heavy air below) which reappears as kinetic energy—winds. Of course, the whirl is continuously being slowed by surface friction thus losing some of its kinetic energy. When, in the last stages of the cyclone's history, there is little further readjustment of air masses and the supply of kinetic energy is cut off, friction gradually brings the giant eddy to a stop.

The net effect of the wave cyclone's history is to disrupt the initial air mass distribution. Part of the cold air mass is swept to lower latitudes near the surface, while some of the warm air mass is transported to higher latitudes aloft.

The sequence of events associated with wave cyclones—as described above—is, of course, an idealization. Few wave cyclones adhere closely to the model throughout their development. However, the model does serve as a guide in furthering understanding of these vortices.

5.6 Tropical Cyclones

Tropical cyclones, as their name implies, are cyclonic storms that are formed over the tropics. In fact, they almost invariably form over the oceans in the latitudes between about 5 degrees and 20 degrees from the equator. They are spawned over all of the tropical oceans except the South Atlantic. Each area of the world has its own local name for this storm, the most common being: *hurricane* (North America), *typhoon* (eastern Asia), *cyclone* (India and Australia), *baguio* (China Sea). Figure 5.16 gives the more common points of origin and paths of tropical cyclones in the world. In the discussion that follows we will use the term which is commonly used in the United States for these storms—hurricane.

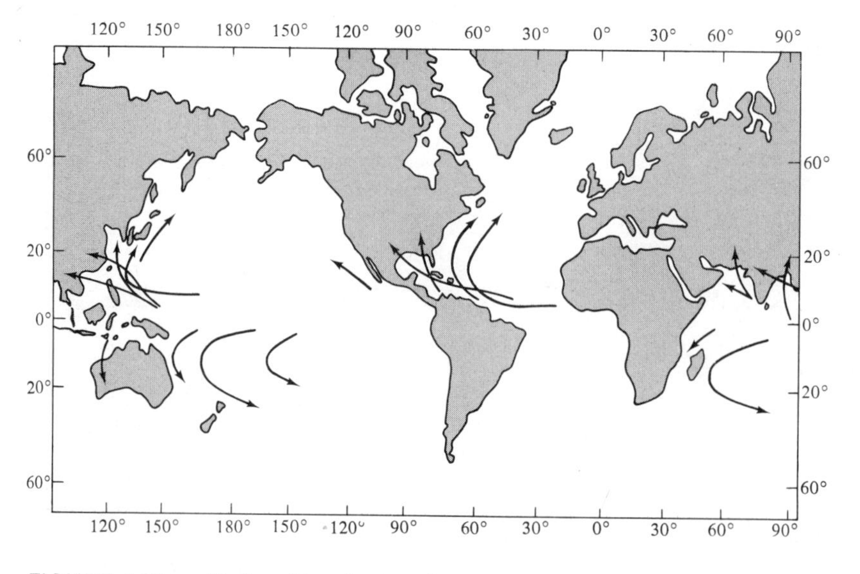

FIGURE 5.16 *Paths of tropical cyclones.*

An astronaut's view of a hurricane, taken from a weather satellite, is shown in Figure 5.17. Note how the bands of clouds spiral in a counterclockwise direction inward toward the center (Northern Hemisphere). From this photograph alone, one could hardly imagine the violence within it. Figure 5.18 presents a picture of the weather conditions in a typical hurricane. (Only a part of the left half is shown, but since hurricanes are approximately circular in form, the conditions on the right would be repeated on the left.) The weather that normally can be expected during the approach and passage of a hurricane can be determined by imagining the observer moving slowly (usually 15-25 km/h) from the outer edge to the center (right to left) and then out to the edge again.

High clouds, which are not too common over the tropical oceans, usually appear 300–500 kilometers in advance of the hurricane. The pressure begins to fall slowly and the winds begin to pick up above the normal 15–30 km/h of the trade winds. Within 300 km of the center, the winds reach gale force (about 50 km/h), steadily increasing in speed, and the pressure begins to fall off a little more rapidly.

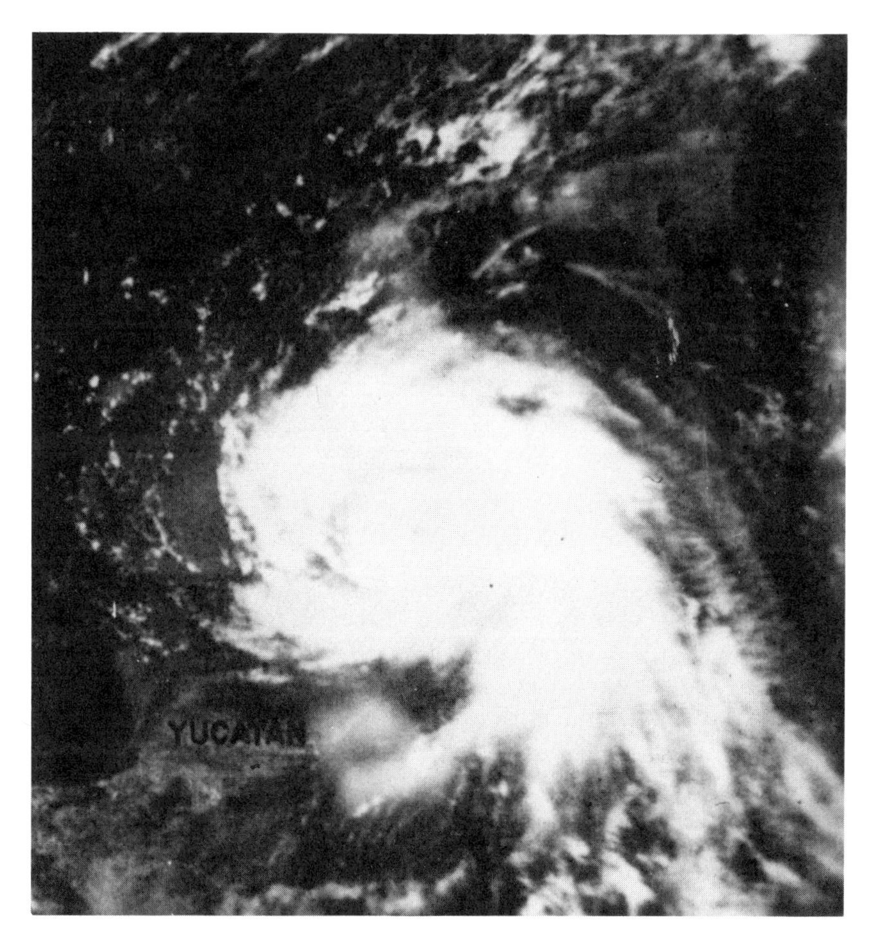

FIGURE 5.17 *Nimbus III satellite photograph of Hurricane Camille, 16 August 1969.*

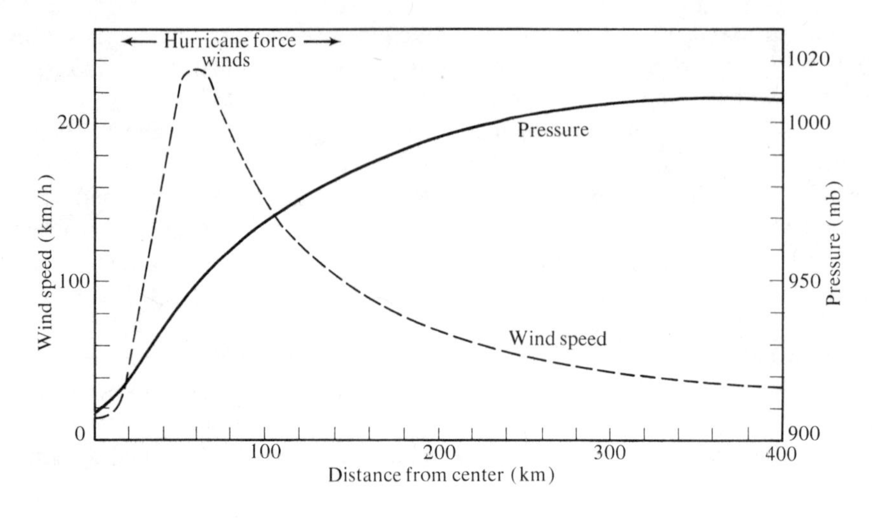

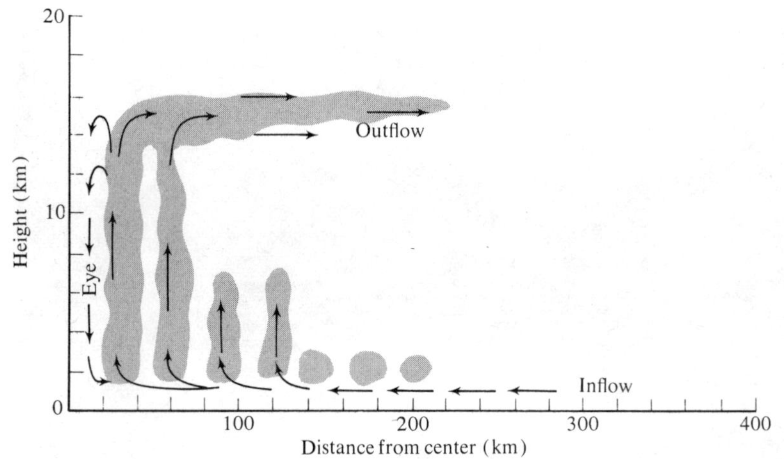

FIGURE 5.18 *Vertical cross section through hurricane and the associated variation of pressure, wind, and precipitation with distance from the center.*

By the time the observer is within 160 km of the center, the winds will be 80 km/h or more, the clouds will be low and menacing, and the pressure will be falling rapidly. Rain usually starts falling 100 or 120 kilometers from the center and increases in intensity until it is coming down in torrents at 30 or 40 kilometers from the center. Winds in this last zone may be as high as 300 km/h.

If the center of the storm passes over the observer, he has a truly startling experience. This center is known as the eye of the hurricane. Quite abruptly, the winds decrease to less than 30 km/h in a distance of 25 kilometers or less. (This distance corresponds to a time interval of less than an hour for the typical hurricane movement.) The rain stops completely, the clouds become thin, and the sun may shine through breaks. The clouds surrounding the eye appear as nearly vertical walls, extending from 1 kilometer to 12 kilometers or more. This is but a respite from the monster storm. Soon, the other half of the "doughnut" will strike and the observer will experience weather conditions similar to those encountered before, except that they will occur in reverse order and the wind direction will be opposite.

The West Indies hurricane season extends from June through November, although most hurricanes occur during August, September, and October. During the seventy-two-year period of 1887 to 1958, a total of 331 hurricanes (an average of 4.6 per year) were reported in the North Atlantic and adjoining waters, in addition to 241 other tropical cyclones that did not reach hurricane intensity (officially, having wind speeds greater than 73 mi/h). About 4 percent of these occurred in the month of June, 6 percent in July, 29 percent in August, 36 percent in September, 19 percent in October, and 3 percent in November. There were only five hurricanes in the other six months of the year. The number of tropical cyclones per year has ranged from as few as two to as many as twenty-one. In the United States, it is customary to identify each season's hurricanes by giving them names in alphabetical succession; thus, the destructive storms of 1963 were Arlene (the first), Cindy (the third), Edith (the fifth), Flora (the sixth and most devastating, causing over 7000 deaths in Haiti and Cuba), Ginny (the seventh), and Helena (the eighth). Until 1978, all of the names of northern hemispheric tropical cyclones were female names; now, however, men's names are also used.

The average lifetime of a West Indies hurricane is nine days, although those occurring during August appear to be more durable, lasting for an average of twelve days. Hurricanes tend to move in the direction of the flow in which they are imbedded, much like an eddy in a river moves downstream. During their early stages, in the Atlantic, while they are still well within the easterly winds, they tend to move toward the west or northwest. If they reach north of about 30 degrees latitude before dissipating, they get caught by the prevailing west winds of the middle latitudes and are swept toward the northeast.

Hurricanes sometimes move in a very erratic fashion. For example, Hurricane Flora (October, 1963) meandered about over eastern Cuba for almost five days and Hurricane Betsy (September, 1965) started toward the northwest over the Bahamas and then passed through the Florida Strait, finally crashing into Louisiana. Although the average speed of hurricanes is about 20 km/h, the speeds of individual storms are extremely variable; when they get caught up in the usual west winds north of 30 degrees latitude, they frequently greatly accelerate, sometimes achieving a speed of over 80 km/h.

The hurricane is a powerhouse of energy. Circulating hundreds of millions of tons of air at speeds of up to 300 km/h or more, the average hurricane generates 300–400 billion kilowatt-hours of energy per day, about 200 times the total electrical power produced in the United States. An average hurricane precipitates 10–20 billion tons of water each day.

The warm, moisture-laden air of the tropical oceans possesses an enormous capacity for heat energy, and it is estimated that most of the energy required to create and sustain a hurricane comes from what is released through condensation. A hurricane is an unusually organized, very large convection system that pumps great amounts of warm, moist air to high levels of the atmosphere at very rapid rates. The arrows in the vertical cross section of Figure 5.18 illustrate the overall convection pattern within a fully-developed hurricane. Warm, moist air rises sharply in the ring between 20 and 60 kilometers of the center. New air flows in toward the center from hundreds of kilometers away. If the air starts out with even a slight counterclockwise rotary motion (caused by the Coriolis force), it will spin faster and faster as it nears the center.

The development of hurricanes occurs through the release of the latent heat of condensation in thunderstorms. Wavelike perturbations in the easterly trade winds have regions of low-level convergence associated with them, somewhat like the perturbations in the westerlies. Under especially favorable conditions, such as very warm ocean temperatures, the thunderstorms may induce a large mesoscale cyclone to form. As the air flows in toward the center of the developing low-pressure system, additional moisture is supplied to the thunderstorms. Thus a feedback is established, with the developing tropical cyclone circulation supplying the thunderstorms with the required moisture and the thunderstorms providing the tropical cyclone with tremendous amounts of latent heat. The warm air produced by the latent heat release is responsible for the extremely low surface pressure. Because the tropical cyclone has a warm core, its intensity diminishes with height. Aloft, the flow becomes anticyclonic.

The greatest damage and loss of life during hurricanes result from flooding of coastal areas by the ocean surges and waves caused by the wind. The sea is in an agitated state hundreds or even thousands of kilometers from the storm center. When wind blows along a water surface it exerts a frictional drag on the water that results in ripples or "waves." The wind drag increases with higher wind speed and so does the size of the waves generated. During a hurricane, air travels at a high speed over long distances, producing waves of great height. The wave heights (vertical distance from crest to valley) often reach 10 meters and sometimes exceed 15 meters in the zone of strong winds. As waves move out from under the winds that generated them, the crests decrease in height and become more regular in shape. Waves of similar height and length between consecutive crests tend to move in groups. These composite waves are known as *swells,* and they can travel thousands of kilometers from the generating area with little loss of energy. When these swells approach a coast, the varying depth to the ocean bottom and the irregularities of the coastline complicate the wave sructure. Sometimes very steep waves travel up estuaries, damaging vessels and piers. If these swells coincide with the normal high tide of the area, they may cause extensive flood damage.

However, the really damaging effects of the wind-churned ocean are not felt until the hurricane center is within a hundred kilometers or so of the coast. Rapid rises in the water level, known as *surges,* result from a piling up of water along the coast by the driving winds. Such "hills" of water can be 5 meters or more above normal sea level. With storm waves riding 10 meters or more above these mounds, large inland areas can be inundated. During a hurricane in 1900, Galveston, Texas, was flooded by just such a surge, which demolished the city and drowned about 5000 persons. In 1961, when Hurricane Carla struck the Gulf coast, a surge was predicted and the affected areas were evacuated beforehand, so there was no loss of human life. In 1969, Hurricane Camille lashed the Mississippi coast with winds up to 300 km/h and storm tides of 6.9 meters, the highest of record. Despite the warnings, about 300 persons lost their lives as the storm moved northeastward over the middle eastern seaboard, and property damage reached $1.4 billion.

When a hurricane moves off the ocean onto land, the frictional drag that the surface of the earth exerts on the wind is greatly increased; however, although the air speed is slowed, it takes a more direct path toward the low center. This more rapid inflow toward the center leads to the gradual dissipation or "filling" of the storm, but at the same time leads to heavier precipitation. Although the genesis of tropical storms is still very difficult to predict and, once formed, their movement is quite erratic, the use of satellites, aircraft, and radar allows the U.S. Weather Service to maintain a close watch of their path and to issue advance warnings of their approach. As a result, in recent years there have been few deaths or injuries. However, property damage is high.

5.7 Tornadoes

The name *tornado* is probably derived from the Spanish word *tornar*, which means "to turn." A tornado is an intense cyclonic vortex in which the air spirals rapidly about a nearly vertical axis. Seen from a distance, it looks like a gray funnel or elephant's trunk extending downward from the base of a cumulonimbus cloud (Figure 5.19). Where this pendant cloud reaches the ground, great masses of dust and debris circle the lowest couple of hundred meters.

The winds associated with tornadoes are too strong to be withstood by the ordinary anemometer, so there are few direct measurements. Estimates from damage to buildings and the impact force of flying objects indicate that speeds range generally between 150 and 500 km/h. Such a wind necessitates a very strong pressure gradient. The pressure drop between the outside and inside of a tornado is usually of the order of 25 millibars, but falls up to 200 millibars have been observed.

The lengths of tornado paths average only about 6 kilometers, but they are extremely erratic. Some touch ground over a distance of only 20 or 30 meters, while others hop and skip over tracks of hundreds of kilometers. Some tornadoes hardly move, while a few have been known to travel at a speed up to 200 km/h. Some last only a fraction of a minute, while others persist for several hours; the average duration is less than 10 minutes. Most move toward the east or northeast (Northern Hemisphere), but every direction of movement has been observed.

During their brief lives, tornadoes can be very destructive. A building in the path of a tornado will certainly be badly damaged, if not destroyed. The cause of the damage to buildings is threefold: the enormous force exerted by the wind, the sudden pressure difference created between the interior and the exterior of the building, and the strong upward air currents. With a rapid pressure drop of 100 millibars, the net outward pressure on the walls of a building could be 200 pounds per square foot, and buildings have been observed to literally explode. Wind pressure can easily reach several hundred pounds per square foot, and powerful updrafts may lift very heavy objects. Many freak occurrences have been reported during tornadoes,

FIGURE 5.19 *Tornado near Enid, Oklahoma. (Courtesy of Leo Ainsworth, National Severe Storms Laboratory.)*

such as showers of frogs that were sucked up from ponds miles away, the "defeathering" of chickens, straws driven through posts, and entire buildings carried for hundreds of meters.

Tornadoes occur infrequently. Although they have been observed in every part of the world outside of the extremely cold regions, they are most common over large continents where strong horizontal temperature contrasts exist—in the United States east of the Rockies, in the southern and middle U.S.S.R., and in southern Australia. In the United States, there are about 150 per year, mostly in the central Plains states. Iowa, Kansas, Arkansas, Oklahoma, and Mississippi have the highest frequency of tornadoes per unit area (Figure 5.20). They occur principally in the afternoon during the spring, but can occur at any time during the day throughout the year.

The formation mechanism of tornadoes is still somewhat obscure. They invariably form in association with severe thunderstorms of the type depicted in Figure 5.7. The rapid updrafts in the thunderstorm produce the evacuation of air and the general lowering of surface pressure. The mesoscale cyclone (Figure 5.7) provides strong horizontal winds with considerable curvature and shear (variation of wind in the horizontal). The actual small-scale tornado funnel is possibly

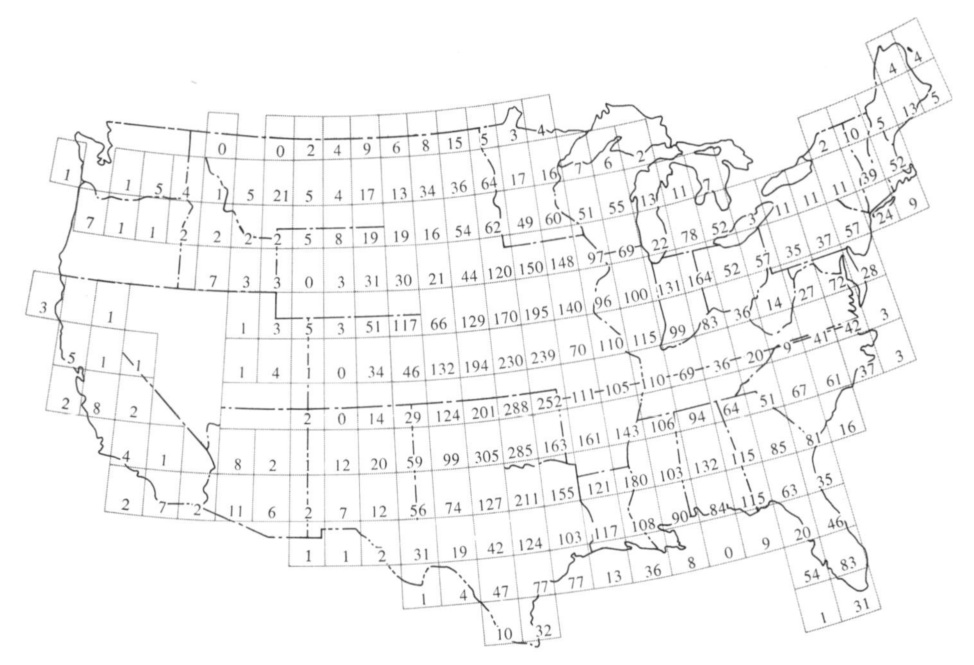

FIGURE 5.20 *Total number of tornadoes by two-degree squares during a 45-year period (1916–1961). (Courtesy NOAA.)*

FIGURE 5.21 *Waterspout over water in Florida Keys. (© Joseph Golden.)*

formed when the mesoscale cyclonic flow becomes unstable and forms smaller-scale, but more intense, vortices. Once formed, strong convection will sustain the vortex until the instability is reduced and friction destroys the whirl.

Tornadoes occasionally form over warm water. Because of the high moisture content of the air, the funnels are heavily laden with water drops so that they look somewhat like a stream of water pouring from the cloud base (Figure 5.21). They are called, for this reason, *waterspouts*. Usually, waterspouts are not as intense as tornadoes over land. Near their base, the winds churn the water surface, producing waves and spray.

A whirlwind that frequently forms on very hot days, especially over deserts, is the *dust devil*. Normally, there are no clouds associated with these and they are no more than a whirling column of dust or sand. They are produced by strong convection near the surface and given a rotation by slight terrain-induced irregularities in the winds. These have been observed to rotate in both senses, clockwise and counterclockwise, with equal frequency.

PROBLEMS

1. Compute the kinetic energy per unit mass, $V^2/2$, for each of the following vortices:
   Tornado: Radius = 0.2 km, speed = 250 km/h
   Hurricane: Radius = 50 km, speed = 150 km/h
   Extratropical cyclone: Radius = 500 km, speed = 50 km/h
   Estimate the total volume of air circulating around each vortex, and from the mean density in the vertical (Appendix 2), calculate the total mass in each circulation system. Compute the total kinetic energy ($mV^2/2$) in kilowatt-hours for each system. Compare the results with the electrical energy production in your city.
2. List the energy sources of each of the major atmospheric vortices. Discuss the theories of how each vortex is initiated. Very few vortices are known to occur over polar regions in winter. Why?
3. During what season and in which hemisphere would you expect the sharpest transition zones (frontal boundaries) between air masses? Why?
4. In which sense does the vortex turn over your bathtub drain? Is the earth's rotation responsible? (Consider the magnitudes of the forces.) Would you expect any rotation if the bathtub were on the equator?
5. Considering the surface winds of the general circulation, by what general route would sailing ships travel from London to New York? By what route would they return? Which of these routes would be the shorter distance?
6. Since the northeast trade winds prevail over most of the Northern Hemisphere between latitudes 10°N and 25°N, why are the winds over southeast Asia from the south during the summer?
7. In a given location, why is the sea breeze usually stronger than the associated land breeze?
8. On the average, the annual temperature range in the Northern Hemisphere is much greater than that in the Southern Hemisphere. Why?
9. An examination of the normal sea level pressure distribution for July shows a significant low-pressure system located over the area of the Mojave Desert, yet almost no precipitation falls in that region during the summer. Why?

Climate

6.1 The Nature of Climate

A common misconception is to think of the climate of a region as the average state of the atmosphere. Not only do the average temperature, precipitation, wind, and other elements determine the climate, but also their variations. The diurnal, day-to-day, and seasonal changes, as well as the extremes of the weather, are important in determining what crops can be grown, how homes and other buildings must be designed, and the way in which many other human activities may be conducted.

The major factors that control climate are, of course, the same that produce weather: (1) the intensity of solar radiation, which is a function of latitude; (2) the reflectivity (albedo) of the earth's surface; (3) the distribution of land and sea; (4) the topography. Many local influences affect the small-scale or *microclimate*—vegetation characteristics, small bodies of water such as lakes, and even human activity that alters the surface properties or the purity of the air.

The task of describing and classifying climate is not an easy one because there are so many facets of the weather that affect human activity. However, the temperature and the water supply are the chief parameters that control the broad-scale distribution of natural and cultivated vegetation. Water supply is dependent on a combination of forces (precipitation, run-off, and evaporation), only two of which—precipitation and evaporation—involve the atmosphere. Evaporation measurements are scanty and unreliable; in any particular area, evaporation depends on the type of surface and vegetation, the temperature of the air, the relative saturation of the air, and the winds. As a result, climatologists have either eliminated evaporation from consideration or have attempted to use temperature alone as an index of evaporation.

6.2 Temperature

The average as well as the diurnal and annual variations of temperature are determined principally by latitude, altitude, and the influence of land and sea. Some of these effects can be seen from the average world isotherms of Figure 3.10. The *latitudinal* variations can perhaps be seen more clearly from Table 6.1.

The mean temperature decreases, and the range increases with latitude. But the difference in the percentage of land mass in the two hemispheres also shows up clearly. The average annual range for the Southern Hemisphere, which not only has less land mass but also has it concentrated in the tropics, is half that of the Northern Hemisphere. The difference between the temperature regimes of stations close to the ocean and those well in the interior of continents is illustrated by Figure 6.1. Note how annual range increases with distance from the ocean shore, especially

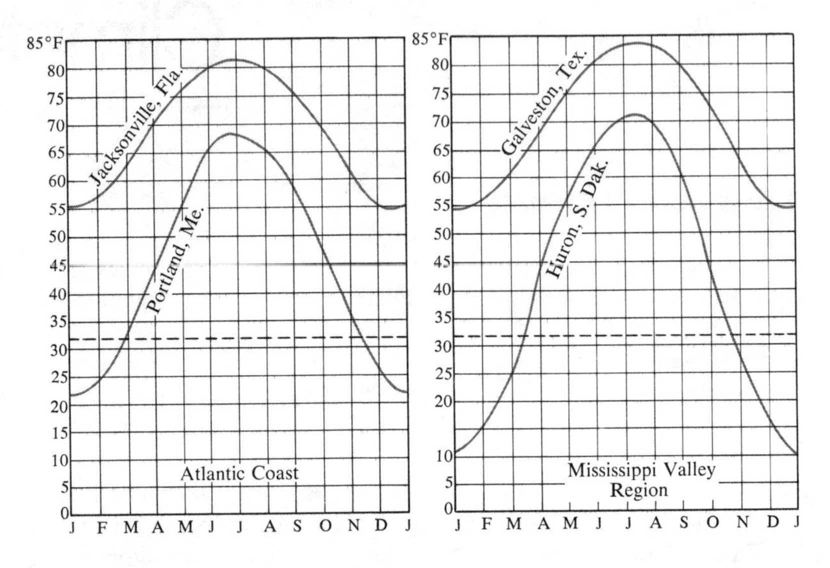

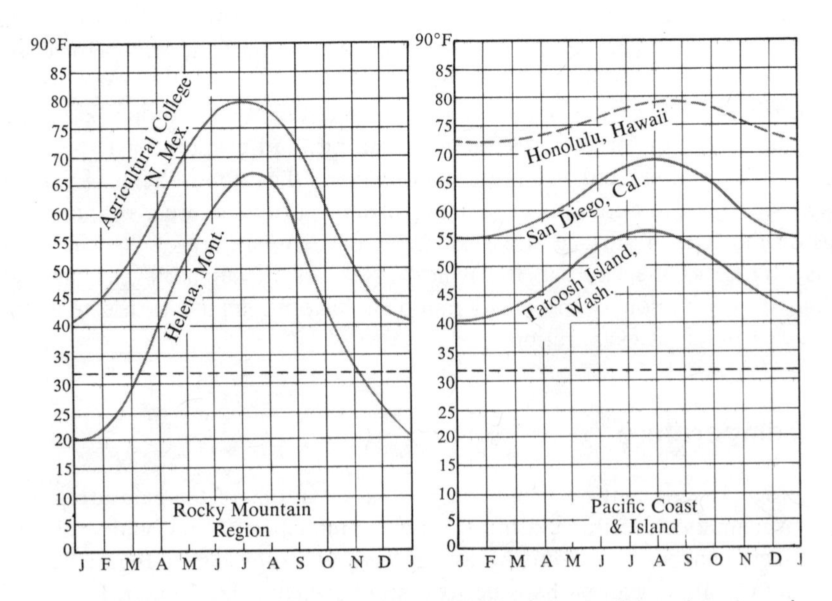

FIGURE 6.1 *Annual temperature variation at continental and marine stations.*

TABLE 6.1 *Mean annual temperature and temperature range and their variation with latitude.*

Latitude	*Mean Temperature (°F)*		*Mean Annual Range (°F)*	
	N. Hem.	*S. Hem.*	*N. Hem.*	*S. Hem.*
90–80°	−8	−5	63	54
80–70	13	10	60	57
70–60	30	27	62	30
60–50	41	42	49	14
50–40	57	53	39	11
40–30	68	65	29	12
30–20	78	73	16	12
20–10	80	78	7	6
10–0	79	79	2	3

from the western shore, because the stations represented are generally within the belt of prevailing westerlies. (Hawaii is an exception.) The range of temperature is an index of what is called the *continentality* of a station. The diurnal temperature variation is also dependent on continentality, as can be seen from Figure 6.2.

The effect of altitude on the temperature range is illustrated by Figure 6.3. In parts of the elevated southwest, the average difference between day and night temperatures in winter is 33°F, while a few hundred miles to the east the range is about a third less. This effect largely reflects differences in moisture and cloudiness in the two areas.

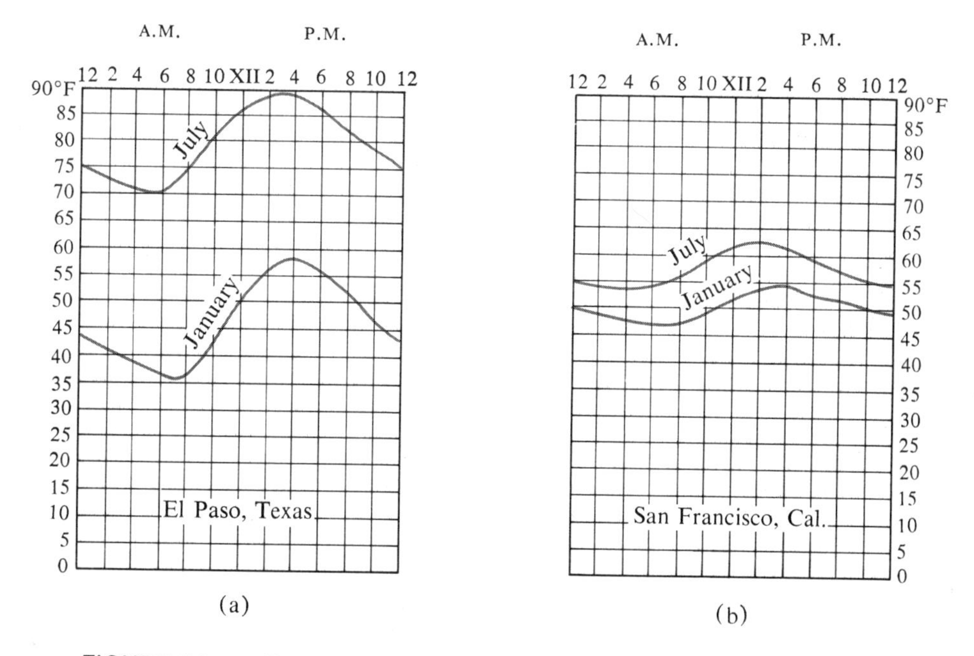

FIGURE 6.2 *Diurnal temperature variations at (a) a continental and (b) a maritime station.*

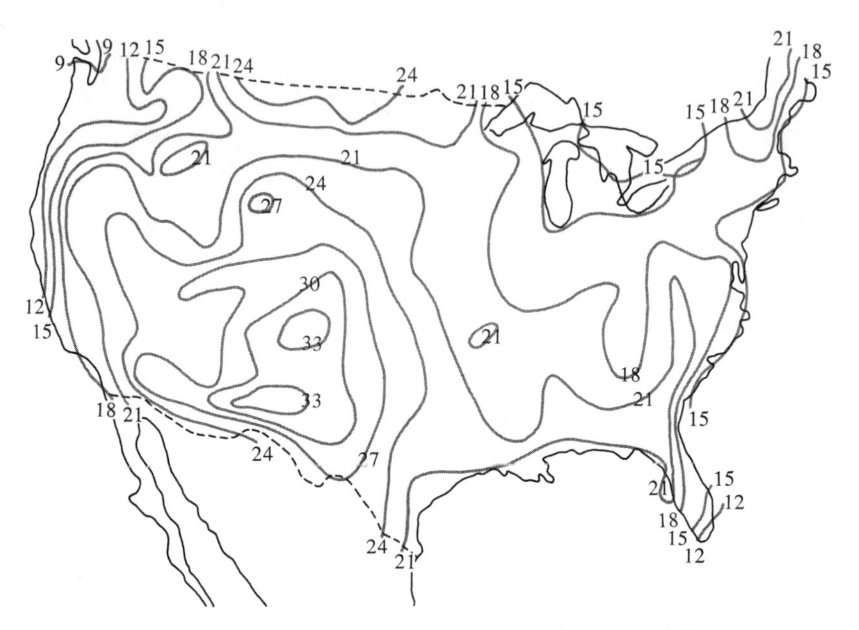

FIGURE 6.3 *Mean diurnal temperature range (January) (°F).*

Temperature inversions Although the temperature typically decreases with elevation in the troposphere, there are special effects of topography and air circulation that cause deviations from the general rule. For example, temperature inversions along the west coast of South and North America at middle and low latitudes are persistent, normal "climatological" features of the temperature distribution. Along the coastal hills, the average temperature is actually slightly warmer at elevations of 1000 meters or so than it is near sea level. Drainage of cold air into low spots in mountainous terrain produces pockets of cold air throughout much of the year. Farmers know this, and they plant their frost-sensitive trees and crops along the slopes, leaving the bottom land for hardier plants.

Temperature indices Temperature is of importance not only to the agriculturist. Human comfort is closely related to the body's heat budget. When the body loses heat faster than it is normally produced, or if the body loses heat more slowly than it should, it suffers discomfort and, under extreme conditions, injury or death. There are several factors that determine the body's rate of heat loss, but one is the temperature of its environment. Some idea of the amount of heating or cooling required in the artificial environments that man creates can be obtained by determining the difference between the mean temperature of a day and some arbitrarily defined ideal temperature (say, 65°F). If one adds up all of these temperature differences over, say, a month, the result can be expressed in terms of *degree days,* a measure of how much heating or cooling will be needed to achieve ideal conditions. Heating engineers find such information useful in estimating fuel and equipment requirements for any locality. You can obtain a rough estimate of the temperature factors which affect your fuel bill from Figure 6.4.

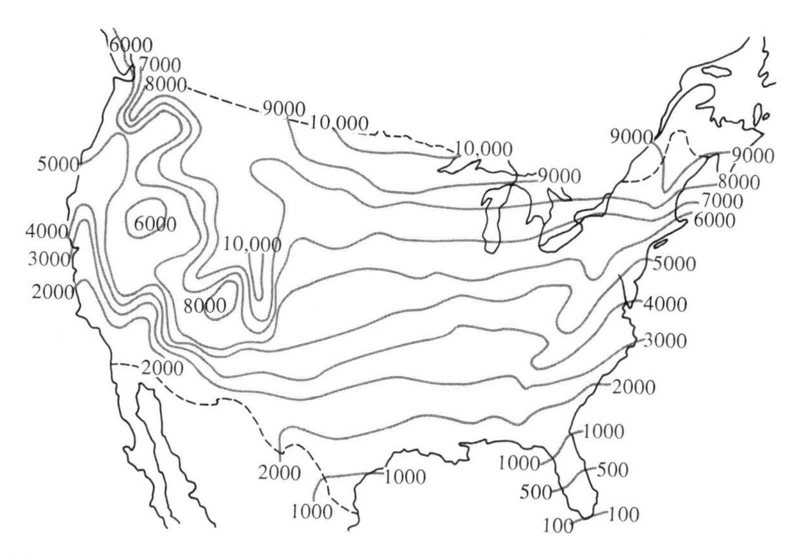

FIGURE 6.4 *Heating degree days over the United States (Base = 65°F).*

6.3 Precipitation

Average cloudiness and precipitation are linked most strongly to the general circulation and topography. Figure 6.5 illustrates the latitudinal variation of precipitation, which conforms, approximately, with the general circulation pattern of Figure 5.2. There is a peak in the doldrums belt where the trade winds converge. The amount

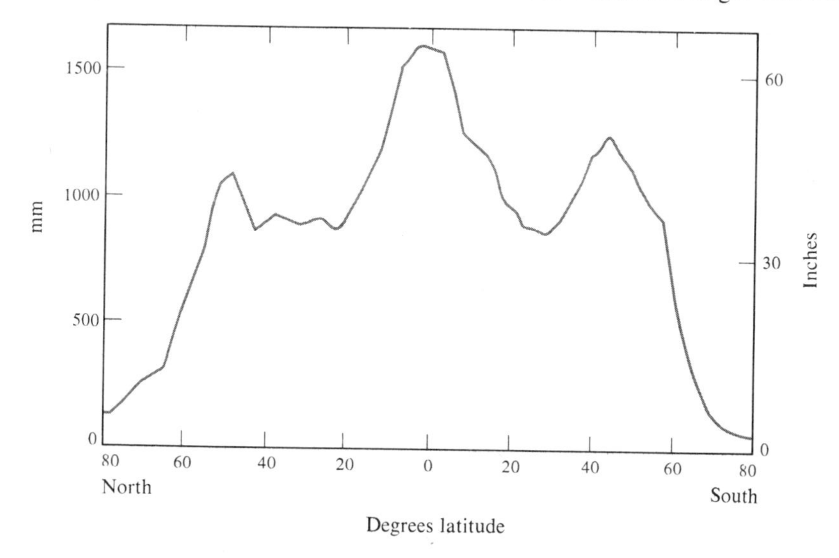

FIGURE 6.5 *Precipitation as a function of latitude.*

drops in the subtropical anticyclone belt, but not drastically. Actually, it is along the eastern edges of the anticyclones in this belt that the great tropical and subtropical deserts are found; the western edges experience upward air motion and ample precipitation. Wave cyclones along the polar front produce much of the rain in the middle latitudes and high altitudes. The polar regions, dominated by anticyclonic flow but experiencing occasional cyclonic weather, are quite arid.

More rain falls on the oceans than on the land (Figure 6.6), and in the belt of prevailing westerly winds, the west coasts of the continents have higher precipitation than the east coasts. Over islands the precipitation is greater than over the surrounding ocean, due to the orographic and convective (heating) effects.

Orographic precipitation (that induced by the forced ascent of air on the windward side of mountain barriers) is also an important factor in the rainfall distribution. Notable examples of orographically produced areas of high precipitation are found on the west side of the Rocky Mountains, on the west side of the Andes in central and southern Chile, and along the west coast of Norway. The rains of the summer monsoon over India and along the southern slopes of the Himalayas are greatly intensified by upslope motion. On the lee side of mountain barriers, there are dry areas, called *rain shadows.* Examples of rain shadows are those found to the east of the Cascades in the states of Washington and Oregon and the arid Patagonia area in Argentina.

The seasonal distribution of rainfall has great significance especially for agriculture. Precipitation is much more useful when it occurs during the growing season of plants than when it occurs at other times of the year. There are a great number of seasonal distributions of rainfall. Some of these are shown in Figure 6.7. Distribution (a) represents the equatorial type. There are two maxima that occur shortly after the equinoxes. These two maxima occur because the intertropical convergence zone oscillates over about 20° of latitude during the year. Each time it passes a point near the equator, rainfall is increased. The other types are: (b) tropical, (c) monsoon, (d) subtropical (west coast), (e) continental, (f) maritime.

6.4 Microclimates

The climate over large areas frequently shows small-scale variations due to the effects of minor topographic features, vegetation characteristics, and even such man-made structures as buildings, roads, and artificial lakes or reservoirs. These microclimatic variations are sometimes very important in determining the crops that can be grown and may affect many aspects of our health and comfort.

The effect that Lake Michigan has on the microclimates in its vicinity is illustrated by Table 6.2. Grand Haven, Michigan, which is generally on the leeward side of the lake, has a higher temperature, more precipitation, more days with snow, and less sunshine during the winter than does Milwaukee, Wisconsin, which is on the windward side. During the summer, on the other hand, Grand Haven has a lower temperature and slightly less precipitation than does Milwaukee.

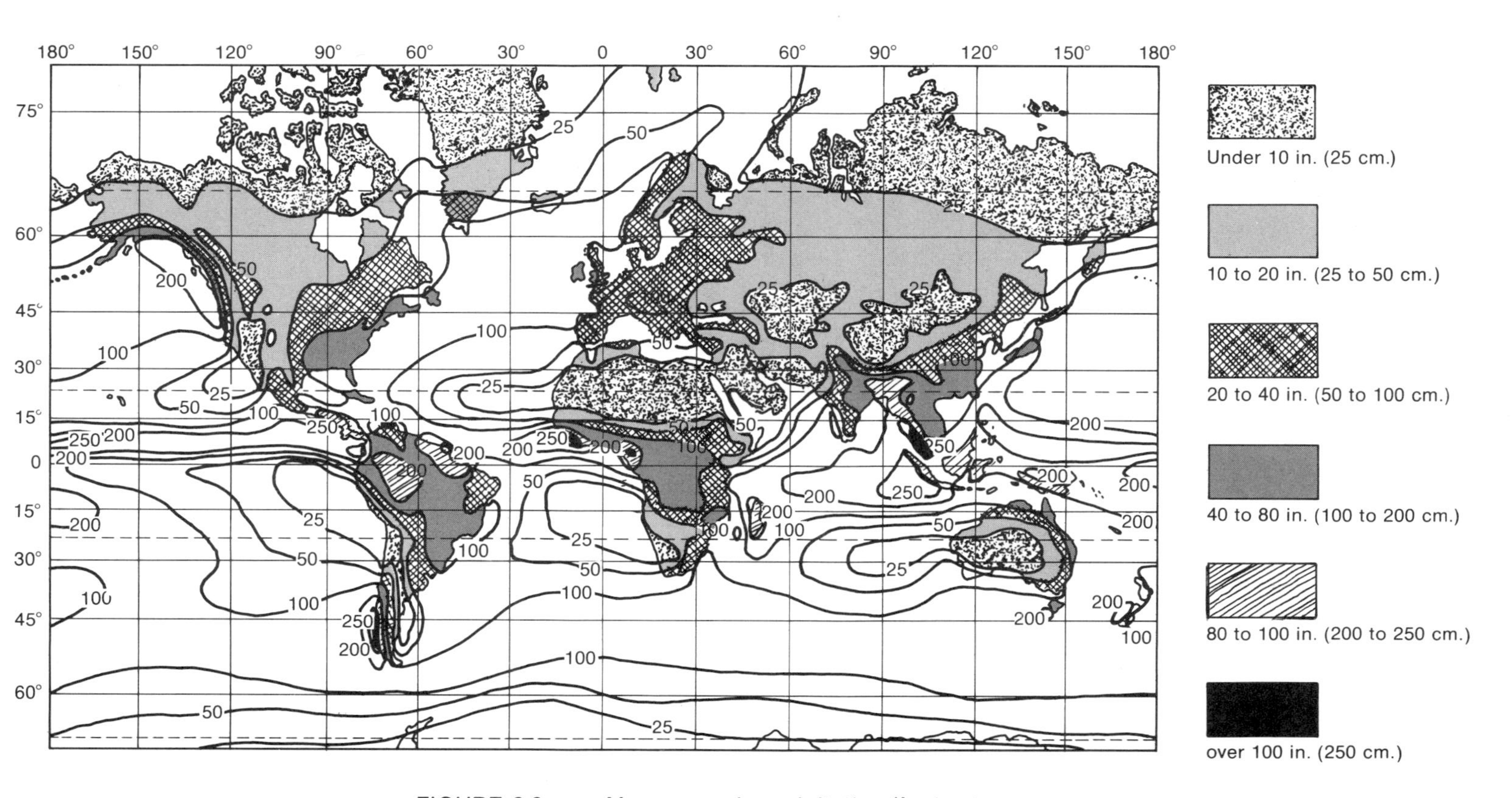

FIGURE 6.6 *Mean annual precipitation (Inches).*

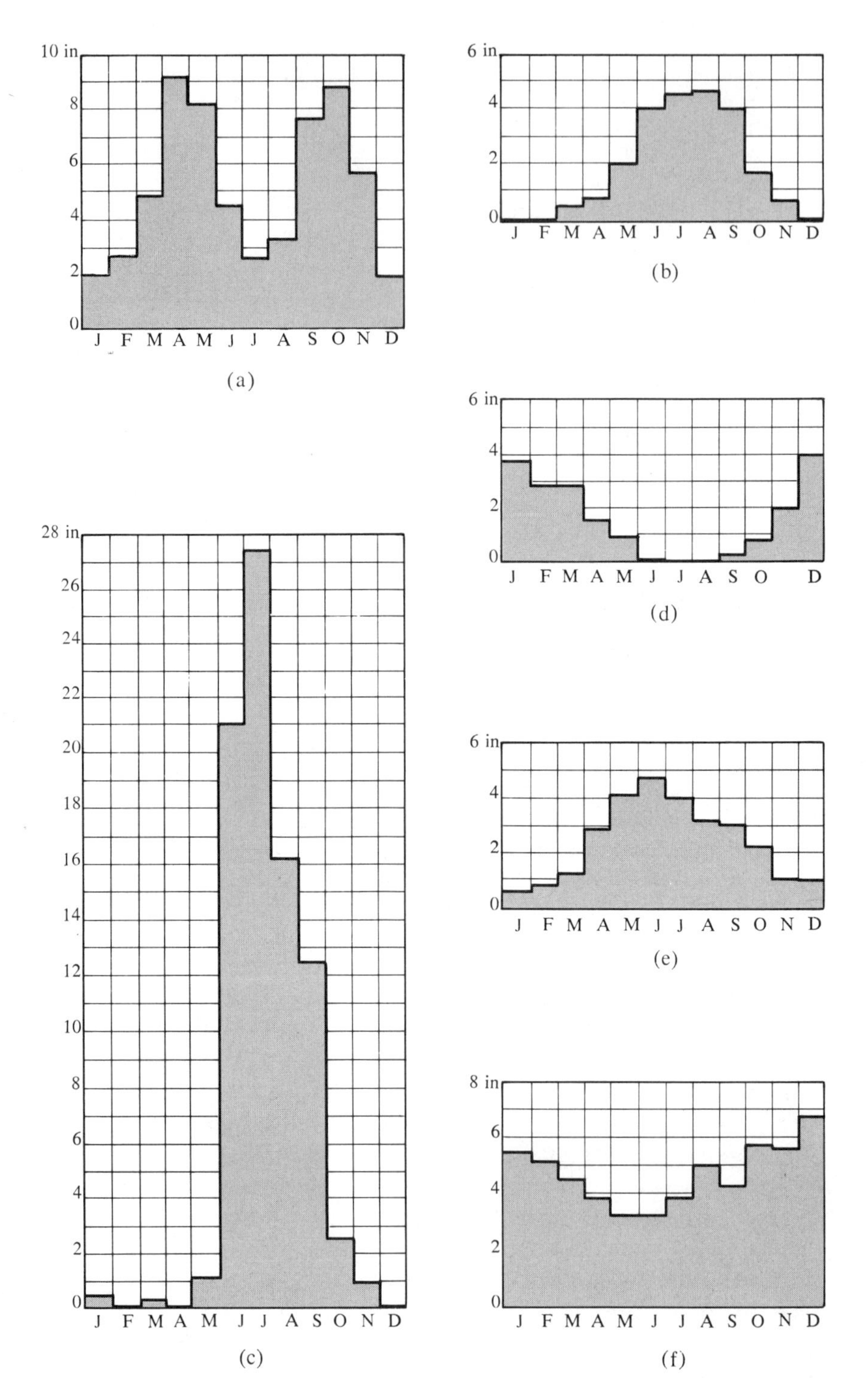

FIGURE 6.7 *Annual distributions of rainfall: (a) Yaounde (Cameroun), (b) Mexico City (Mexico), (c) Bombay (India), (d) Sacramento (California), (e) Omaha (Nebraska), (f) Valencia (Eire).*

Trees and plants produce microclimatic variations primarily because of their effects on the moisture supply and the wind flow near the surface. Transpiration from vegetation causes locally higher humidity, while the soil tends to inhibit precipitation runoff. Temperatures and wind speeds are lower within forested areas than in the open.

TABLE 6.2 *Microclimatic Difference between Windward and Leeward Shores of Lake Michigan (After Landsberg)*

Climatic element	*Time of year*	*Milwaukee (windward shore)*	*Grand Haven (leeward shore)*
Mean temperature	January		3.6°F higher
Mean temperature	August	2.0°F higher	
Precipitation	Dec.–Feb.		2.09 in. more
Precipitation	June–Aug.	0.55 in. more	
Snowfall	Jan.–Dec.		44 days more
Sunshine	Dec.–Feb.	20% more	

With continued proliferation of industrial and urban development, the importance of the microclimatic environment increases. The climate of large cities is characterized by increased "smog," dust, and waste gases from traffic, industrial sources, and domestic heating. While these pollutants prevent some of the sun's radiation from reaching the city itself, the decreased solar heating is more than compensated for by absorption and reradiation of the sun's energy by streets and walls of buildings, and by the lack of evaporative cooling, which would be present in forests and fields. Furthermore, the buildings of the city tend to impede the natural flow of air. Consequently, the city center tends to be warmer than its surroundings and to suffer considerably more from impurities in the air. Recent studies also suggest that cities tend to have more precipitation and fog than the surrounding countryside.

By making use of his knowledge of microclimates, man is able to accomplish certain desirable changes in the environment. For example, removal of a thick cushion of moss was undertaken in the Yakut area of Siberia during World War II. Since the moss served as a heat insulator and also caused considerable evaporative heat loss, its removal permitted increased absorption of solar radiation by the ground. Consequently, the heretofore frozen soil thawed to a sufficient depth so that crops could be grown. Other examples of the use of microclimatic knowledge include the planting of rows of tall trees as windbreaks in the valleys of California, and the designing of homes to take advantage of existing favorable microclimates produced by existing hills, streams, and vegetation in the local area.

Climatic extremes Of considerable interest to almost everyone is "record-breaking" weather, and newspaper headlines periodically note the occurrence of a record low temperature or an excessively heavy snowfall at some location almost every year. Table 6.3 shows the highest and lowest values of a number of weather

TABLE 6.3 *Climatic Extremes—Highest and Lowest Observed Values of Certain Meteorological Elements*

World records	*U.S.A. Records*
Temperature—Highest	
136°F at Azizia, Libya, September 13, 1922	134°F in Death Valley, Calif., July 10, 1913
Temperature—Lowest	
−127°F at Vostok, Antarctica, August 24, 1960	−80°F at Prospect Creek, Alaska, January 23, 1971
Rainfall—Greatest Annual	
1041.78 in. at Cherrapunji, India, August, 1860–July, 1861	578.00 in. at Puukukui, Maui, 1950 269.30 in. at Little Port Walter, Alaska, 1943 184.56 in. at Wynoochee Oxbow, Wash., 1931
Rainfall—Greatest Monthly	
366.14 in. at Cherrapunji, India, July, 1861	107.00 in. at Puukukui, Maui, March, 1942 71.54 in. at Helen Mine, Calif., January, 1909
Rainfall—Greatest 24-hour	
73.62 in. at Cilaos, La Reunion, Indian Ocean, March 15–16, 1952	38.70 in. at Yankeetown, Florida, September 5–6, 1950
Longest Period without Rainfall	
19 years at Wadi Haifa, Sudan (entire period of record)	767 days at Bagdad, Calif., October 3, 1912–November 8, 1914
Snowfall—Greatest 24-hour	
	76 in. at Silver Lake, Colorado, April 14–15, 1921
Snowfall—Greatest Annual	
	1027 in. at Paradise Ranger Station, Washington, 1970–1971
Highest Wind Speed at Surface	
231 mi/h at Mt. Washington, N.H., April 12, 1934	231 mi/h at Mt. Washington, N.H., April 12, 1934
Largest Recorded Hailstone	
	1.67 pounds at Coffeyville, Kansas, September 3, 1970

elements, representing extremes up to the current time. However, it is axiomatic that, by their very nature as extreme weather conditions, such records will continue to be broken. Probably by the time this text is published, at least one of these records will have been exceeded.

6.5 Climatic Change

Weather is noted for its variability. However, climate also changes. If one were to plot the value of some element of weather, such as temperature, as a function of time, fluctuations over all time scales would appear. For example, if the air temperature were averaged over hourly intervals, these averages would change from hour to hour in response to the earth's daily rotation. If these 24 hourly temperatures were then averaged for each day, one would notice day-to-day oscillations resulting from air-mass changes. Similarly, a distinct oscillation during the year is to be expected because of the earth's revolution around the sun. But if one were to average all the daily temperatures over each year, year-to-year fluctuations would appear, presumably due to changes in the general circulation of the atmosphere. Even averages computed over decades and centuries fluctuate, as do averages over thousands and millions of years, judging from the indirect evidence available. Thus, weather is not a constant on any scale of time, but rather it appears to vary over all possible periods.

These cycles or periods (intervals of time over which the weather elements repeat themselves) are not generally very regular. Even the most regular of them—the daily and annual variations—are not unchanging. For example, the start and end of the summer monsoon over Southeast Asia vary considerably from year to year. In fact, forecasting from assumed periodicities in the weather has never been successful precisely because of the great irregularity of such cycles, even for periods much less than a year.

Systematic weather observations do not exist for longer than about 300 years, so long-period changes of climate must be inferred from evidence other than direct measurement. Most of such indirect evidence gives information only on precipitation and temperature. First of all, there are written records, which permit extrapolation backward in time for a few thousand years. Then there are the varying widths of growth rings of old trees, the migrations of people, the fluctuations in levels of lakes and rivers, and the succession of plant types, which provide clues to the changes that have occurred during the past ten to twenty thousand years.

Geological evidence must be used to extend the time scale further into the past. For this purpose, the types of flora and fauna in fossilized form found in an area are indicative of the climatic characteristics at the time they lived. The advance and retreat of glaciers (snow fields), which leave their imprint on the earth that they have traversed, are signs of changing climate.

Very little is known about conditions before 600 million years ago (before the Cambrian period), but great "ice ages" have evidently occurred at intervals of about 250 million years: one some 700 to 1000 million years ago, another about 550 million years ago, a third 250 million years ago, and the latest, and best documented, which began roughly a million years ago and has only recently (geologically speaking) ended.

There have been numerous advances and retreats of the ice sheets. For example, in the last great Quaternary Ice Ages, there have been at least four major advances and retreats of the ice. In the last of these advances, which lasted about 100,000 years, great ice packs extended hundreds of miles south of the Canadian border into the Mississippi Valley. Emergence from this last ice age began less than 20,000 years ago, with its completion only some 8000 years ago. At the peak of this glaciation, reindeer and musk-oxen penetrated the central United States, and walrus were found along the Georgia coast.

Beginning about 10,000 years ago, the European climate increased in warmth and dryness compared to the coldness and wetness of the preceding periods; between about 6000 and 3000 B.C., the climate reached an optimum—warm and humid—with annual mean temperatures about 2°C higher than today and with year-round rain. The European climate cooled somewhat and was rather dry between 3000 B.C. and 500 B.C. It cooled markedly between 500 B.C. and A.D. 200; during this period, the winters were snowy and cold and the summers cool and wet. Although the worldwide trend over the past 10,000 years has been general warming with receding glaciers, during the past thirty years, the mean temperature has decreased, with the inevitable speculation that the trend may be reversing.

The evidence for climatic fluctuations over all scales of time is fairly conclusive. It has been shown in earlier chapters of this book that there are numerous factors that play roles in determining the state of the atmosphere: the energy received from the sun and its latitudinal distribution, the transparency of the atmosphere to solar and terrestrial radiation, mountain barriers, the frictional drag of the earth's surface. Just as there exists a variety of causes for daily, day-to-day, and week-to-week changes of weather, so too it appears that there must exist changes in one or more of these factors that were responsible for the longer period, although irregular, changes of the earth's climates. The possible causes for these long-period climatic variations can be grouped into three broad categories: 1) *astronomical* (anything that alters the amount, type, or distribution of solar energy intercepted by the earth); (2) *atmospheric transparency* (anything that changes the the transmission of radiation through the atmosphere); and (3) *earth's surface* (anything that alters the energy flow at the earth's surface or its geographic distribution). Some postulated causes for long-period climatic changes within each group are listed in Table 6.4.

It should be emphasized that no single factor is likely to be responsible for *all* of the observed changes in climate. All causes have had *some* effect; the problem is to determine which are most significant. Specific causes have been postulated to cover almost every observed period of fluctuation. We shall comment on only a few of the most seriously considered theories.

TABLE 6.4 *Possible Causes for Climatic Change*

Cause	*Approximate Range of Periods Induced (years)*
I. Astronomical changes:	
A. Solar aging	10^9
B. Passage of solar system through galactic dust	10^8–10^9
C. Solar output variability	10^1–10^8 (?)
D. Earth orbit changes	10^4–10^5
II. Atmospheric transparency changes:	
A. Volcanic dust in the stratosphere	10^0–10^8 (?)
B. Carbon dioxide content changes due to natural causes	10^4–10^8
C. Carbon dioxide content changes due to recent industrialization	10^1–10^2
D. Changes of other gaseous constituents	10^8–10^9
E. Dust particles introduced by man's activities	10^0–10^2
III. Earth's surface changes:	
A. Migration of the poles	10^7–10^9
B. Continental drift	10^7–10^9
C. Lifting of mountains and continents	10^7–10^9
D. Relative sizes of ice caps and oceans	10^4–10^5 (?)
E. Slow ocean circulation from great depths	10^3–10^6 (?)
F. Slow adjustments between atmosphere and ocean	10^0–10^3

Astronomical

One theory of this group is that the energy output of the sun, either in terms of quantity or spectral distribution, varies with time. Measurements over the past decades indicate that any variations in the *total* energy output are probably smaller than the accuracy of the measurements, which is about ± 2 percent. However, there are intermittent outbursts of corpuscles (charged particles) and very short wavelengths (ultraviolet) of radiation associated with disturbances on the sun that are largely absorbed in the outermost layers of the atmosphere. These outbursts are known to affect the ionosphere and the earth's magnetic field and to produce auroras, but it is difficult to understand at the present time how these very short waves of radiation, absorbed by the very thin atmosphere above 30 or 40 kilometers, can appreciably affect the great mass (more than 99 percent) of the atmosphere that lies below, where the weather occurs.

These outbursts from the sun are prevalent during years of relatively high frequency of sunspots. Sunspots have been observed and counted since the days of Galileo. Well-defined oscillations in the number of spots appearing on the face of the sun occur over periods of about 11 and 23 years. Although periodicities of about the same lengths have been found for certain elements of weather, such as air circulation, pressure, and temperature, it is difficult to say whether such correspondence is merely accidental without knowing how the phenomena might be physically linked.

Just how changes in solar output might affect the climate is uncertain. It would appear unlikely that the effect would be merely to increase or decrease the average world temperature. Rather, an increase in solar output would result in greater heating in the equatorial regions than in the polar regions, thus increasing the latitudinal temperature gradient and therefore intensifying the atmospheric circulation. Stronger winds would lead, in turn, to increased evaporation, increased precipitation, and more extensive glaciation. Conversely, a decrease in solar output would reduce the temperature differences between poles and equator, thus weakening the atmospheric circulation, which in turn would lead to diminished glaciation. Very little is known about the variation in solar output during the earth's lifetime of about 5 billion years. Until there is some direct evidence of abrupt changes in the sun's energy output, it will be difficult to either accept or discount this theory as a significant cause of climatic change.

A second theory in this group of "astronomical" causes regards earth orbit changes. There are three types of variations in the earth's motions: (1) The obliquity of the ecliptic (i.e., the angle between the plane of the earth's orbit and the equatorial plane), which is now about 23½ degrees, has changed by about 2½ degrees in 45,000 years. When the angle is large, the seasons are more extreme and the pole–equator temperature difference in winter is increased. (2) Perihelion (the point in the earth's orbit where the earth is closest to the sun) now occurs during early January, but the date of its occurrence advances through the year with a period of about 21,000 years. (Thus in 10,000 years, perihelion will occur in July.) When winter in one hemisphere coincides with perihelion, it will be a little milder than when it coincides with aphelion. This variation is called the precession of the equinox. (3) The eccentricity (*ellipticity*) of the earth's orbit changes over a period of about 85,000 years. Although the difference in radiation received from the sun at perihelion compared to that received at aphelion is now only 7 percent, with maximum eccentricity the difference is as much as 20 percent.

Although these orbital changes only slightly affect the average yearly radiation received by the earth, by changing the seasonal distribution, they may significantly alter the summertime melting and shrinkage of ice caps. Indirect evidence from ocean deposits confirms that orbital changes are significant producers of climatic change; however, they can hardly stand alone, since they cannot explain either the relatively short-period variations (less than 20,000 years) that have been observed or the long period without an ice age that occurred just before the last million years.

Atmospheric transparency

Volcanic eruptions sometimes spew fine dust particles into the upper atmosphere, and the particles can persist for years. These dust particles deplete some of the sun's rays by increasing the amount that is scattered by the atmosphere. In other words, increased dust causes an increase of the earth's reflectivity (albedo). For example, a mere 1 percent increase in the earth's albedo could lower the earth's mean temperature about 1½ °C. Volcanic dust, trapped in the lower stratosphere, has been cited by some scientists as the cause of notable drops in the mean world temperature that have been observed in certain years during the past two centuries.

Dust counts at several places around the world indicate that the dust content of the atmosphere has been increasing steadily during the past few decades, perhaps due not only to volcanic eruptions but also to human activity. The mean world temperature has dropped by about 0.3 °F in the past 25 years.

In addition, a variety of gases are given off by volcanoes, including carbon dioxide, which is a good absorber of infrared radiation in certain wavelengths. But there is some doubt as to whether an increased amount of carbon dioxide in the atmosphere would materially increase the absorption of infrared since the average concentration is already quite effective. Also, the radiation absorption bands of carbon dioxide overlap those of water vapor. There is also some question as to whether the oceans regulate the concentration of carbon dioxide by absorbing any excess. Measurements of carbon dioxide content in the atmosphere are as yet too few to determine whether there are any long-term variations. There is evidence that the concentration of carbon dioxide in the air over cities is greater than in the countryside, and it may be that the increased consumption of fossil fuels, such as coal and oil, during the past couple of centuries has produced an increase of carbon dioxide in the air and, therefore, higher temperatures over urban areas.

Earth's surface

Anything that might alter either the properties of the earth's surface or their distribution over the globe could lead to climatic change. Evidence from magnetic data in rocks (*paleomagnetism*) has shown that there may exist a slow wandering of the positions of the poles on the earth's surface; i.e., in the past, the poles may have been at geographic points other than their present locations. The difficulty with this fact as an explanation of climatic change is that different areas would be affected at different times, but present evidence is that glaciers have occurred almost simultaneously over many regions of the earth.

Continental drift across the globe, possibly induced by convective currents within the earth's mantle, would certainly produce large changes in climate, but here again, they would not be simultaneous over many different land areas.

Mountain-building is generally considered to be an important cause of climatic changes, especially those taking several hundred thousand years or more. When a mountain is created, the increased elevation itself gives rise to lower temperatures, but in addition, it reduces the exchange between polar and equatorial regions. (For example, the Himalayas are quite effective in preventing the mixture of warm and cold air masses to their south and north.) Mountain-building and erosion take a long time and cannot explain the shorter periods of climatic change.

Expansion and contraction of the great polar ice caps would certainly have an important effect on world climates, principally because ice is such a good reflector (poor absorber) of solar radiation. Thus an increase in the area of the ice caps would decrease the total amount of solar energy absorbed by the earth. It has been postulated, for example, that the great ice pack over Antarctica may periodically (about every 70,000 years?) slip into the adjoining oceans, spreading out to form continent-sized shelves. This slippage may occur when the ice depth over the Antarctic continent reaches some critical value such that the ice at the bottom, which is under enormous pressure, begins to melt. (Heat flowing from the earth's interior may help to melt this bottom ice.) If the ice were to extend itself over millions of square kilometers of ocean, the mean world temperature might fall several degrees centigrade, and sea level would rise some tens of meters.

The oceans represent a tremendous reservoir of heat. Recently, it has been shown that the cold water at great depths may slowly rise and mix with surface water. These deep water circulations may take tens of thousands of years to complete. Since the oceans cover such a large proportion of the earth's surface, they are, of course, an important source of heat in the atmosphere; thus changes in the ocean's surface temperature could result in significant changes of temperature and moisture in the atmosphere. However, it seems unlikely that such circulations could, by themselves, be responsible for the magnitude of temperature change observed during ice ages.

In summary, there are so many factors that control climate that it is little wonder that postulated causes for climatic change are so numerous. It would appear likely that more than one of the causes mentioned above is responsible for climatic changes, just as the daily weather is the result of many factors at work. But even a change in a single factor is likely to produce a complex reaction by the atmosphere. For example, an increase in radiation from the sun might cause this chain of events: the mean temperature is raised as well as the north-south temperature gradient; this may lead to increased air circulation and, therefore, greater convection and cloudiness; the clouds might then cut off some of the radiation. Thus the atmosphere may have a sort of "built-in thermostatic control."

Many of the theories that have been presented could produce some climatic change and perhaps did. The question that investigators try to answer is, Which is the most significant cause of the observed climatic changes? At the moment, popular theories for climatic change include variations in the earth's orbital characteristics, mountain-building, and changing solar energy output. Until meteorologists can develop better models of atmospheric processes and of the general circulation, and until geologists are able to more accurately date the climatic changes of the earth, the causes and prediction of the future climate can at best be highly speculative.

PROBLEMS

1. Why is there so much snow and ice over the Arctic and Antarctic even though precipitation is light? Assuming the average precipitation over the Arctic is that of the average at latitude 80°N, what would be the minimum age of the ice at the bottom of a 30-meter iceberg? (Neglect compression of the snow.)
2. Why does the equatorial type of rainfall distribution have two maxima that occur shortly after the equinoxes?
3. Why does the maximum of precipitation occur in summer over the interiors of continents?
4. Explain the maxima of thunderstorms in Florida and over the Rockies (Figure 5.10).
5. From a knowledge of the general circulation, deduce the characteristics of the climate of the state of Washington, taking into account the topography.
6. Why are there relative minima of precipitation at latitudes 20°–30°S and N and maxima at 40°–50°S and N?
7. Suppose the obliquity of the ecliptic decreased from 23½° to 10°. How would the seasons be affected? When would the winter solstice occur? How would the length of a June day in New York City be changed? A November day?

7 Weather Forecasting

Accurate prediction is the goal of all scientists. But few physical scientists have a more complex, more frustrating, or more challenging medium to work with than the meteorologist. It has been pointed out in previous chapters that circulations of almost all sizes exist in the atmosphere, that the earth's surface is not only "corrugated" but is covered with different materials, and that even the constituents of the air, especially that very important one, water vapor, vary considerably in both space and time. With such intricate and ever-changing weather patterns that can be only very inadequately observed, it is little wonder that improvements in the accuracy of prediction have been painfully slow.

7.1 Weather Map Analysis

Weather forecasting starts with atmospheric observations. More than 10,000 land stations and hundreds of ships take regular "surface" observations, and more than 100 land and sea stations make regular radiosonde observations. In addition, aircraft take observations over ocean areas where ship reports are scarce, especially in zones where hurricanes form. Through international accord, all the nations of the world exchange their information, except in time of war. To facilitate communication a standard weather reporting code has been adopted.

After the forecast offices of each country receive the data, the first step is to prepare a three-dimensional analysis of the atmosphere. There are many techniques for studying the variation of atmospheric properties in both the horizontal and vertical, but the most widely used method is the construction of a series of charts that represent horizontal cuts of the atmosphere from sea level to above the tropopause. The *sea-level chart* is by far the most complete, both in terms of the number of stations reporting and the number of variables represented at each point. This is the familiar weather map that appears in many newspapers. Examples of surface weather maps are given in Figure 7.1. At the location of each reporting site, its complete observation is plotted, following the model shown in Appendix 4. A vast amount of information is packed into the surface data: a complete description of the clouds, temperature, humidity, pressure, wind, precipitation, and restrictions to visibility. The weather chart analyst's first step is to draw isobars so that he can determine the position, size, and intensity of high- and low-pressure centers. His next step is to locate fronts; this he does by noting the hori-

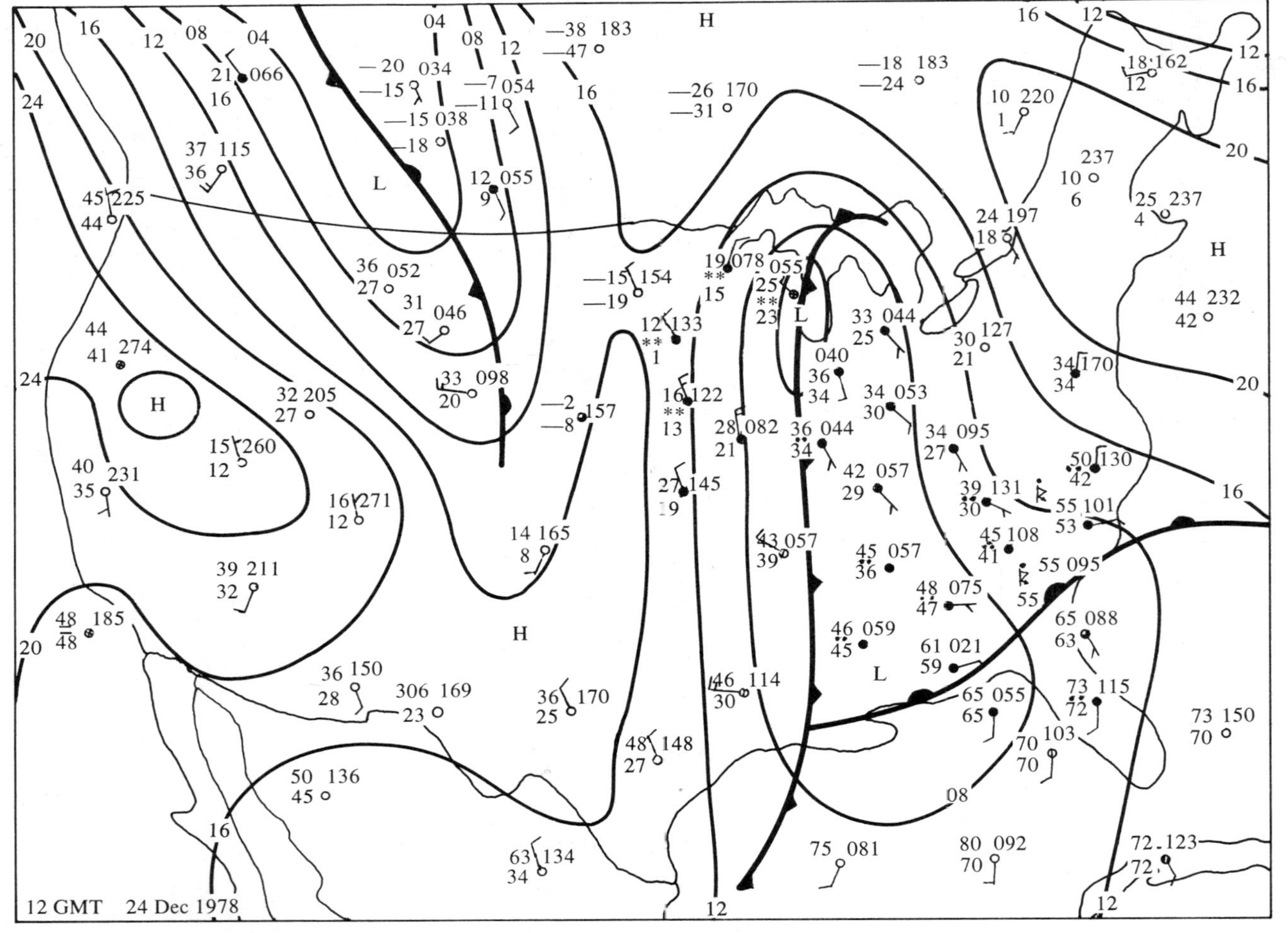

FIGURE 7.1(a) *Surface weather map for 12 GMT December 24, 1978.*

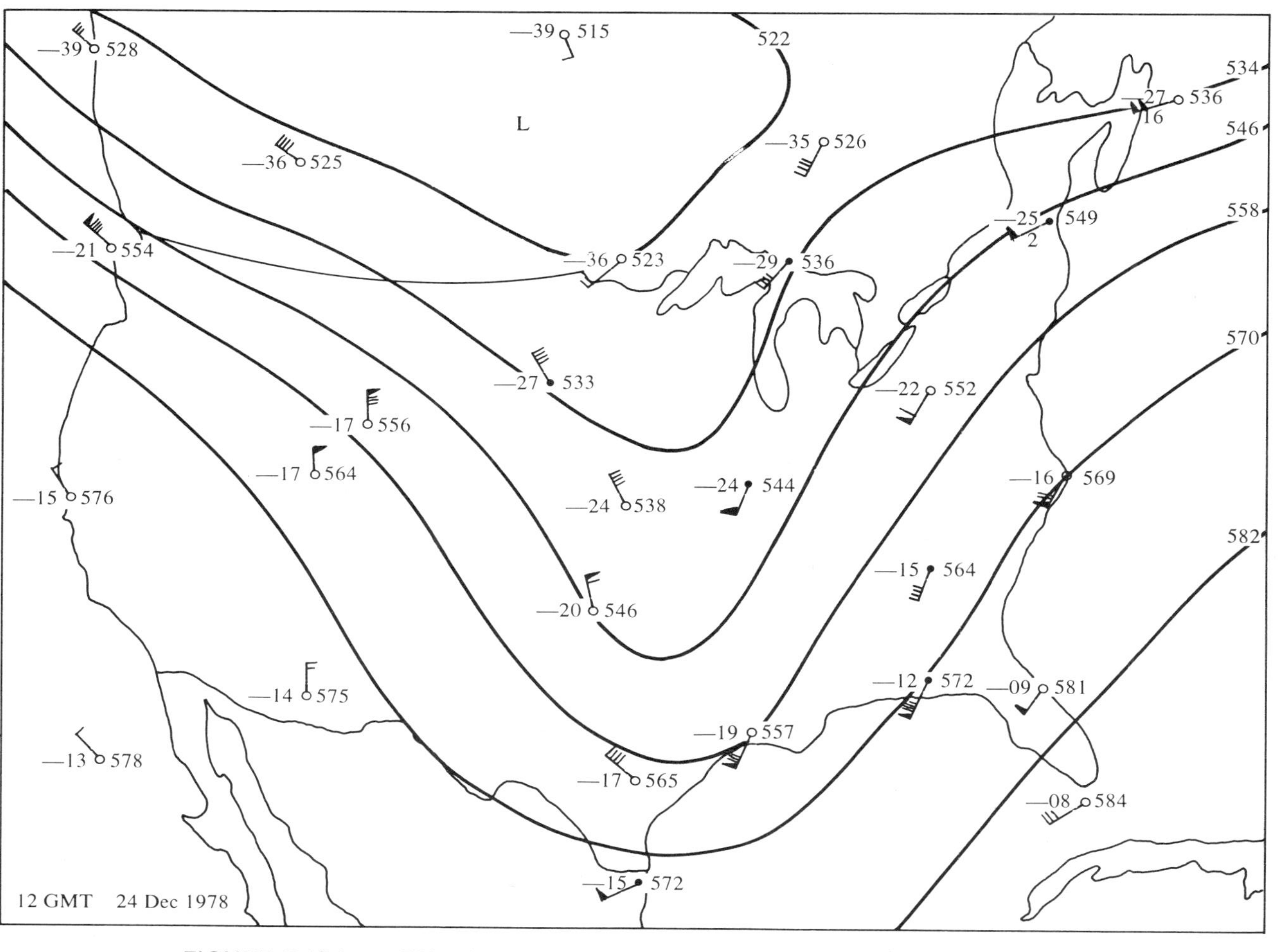

FIGURE 7.1(b) *500-mb map for 12 GMT December 24, 1978.*

zontal distribution of temperature, humidity, pressure, wind, cloudiness, and precipitation, taking into account the polar front model discussed in Chapter 5. He will also confirm his surface analysis by a study of the temperature, humidity, and wind flow aloft.

Conditions aloft are usually represented by charts of isobaric (constant pressure) surfaces. These are equivalent to charts of constant altitude, since the horizontal pressure gradient at a constant level is proportional to the slope of an isobaric surface (Figure 4.2). On constant pressure charts, the wind conforms to the gradient of contours in the same way that the wind is determined by isobars on a constant altitude chart. The plotting model used for these "upper-air" charts is given in Appendix 4.

Examples of analyzed upper-air charts (500 millibars, approximately 5500 meters) are given in Figures 7.1(b)–7.3(b). The constant pressure surfaces that are routinely analyzed are 850, 700, 500, 300, 200, and 100 millibars (approximate altitudes of 1500, 3000, 5500, 9200, 11,800, and 16,200 meters).

Many other types of analyses are regularly made at most meteorological forecast centers—vertical cross sections of the atmosphere showing the distribution of temperatures, winds, and moisture in areas of particular concern; analyses of wind speed that help to locate the jet stream; analyses of atmospheric stability for the prediction of thunderstorms and tornadoes; and others that describe the properties of motion (e.g., the air's "spin"), potential energy, etc.

7.2 Forecasting Techniques

After the current state of the atmosphere has been determined by analyses such as those described in the previous section, the next very difficult step is to determine the future patterns of the meteorological variables (temperature, wind, etc.). Several techniques have been employed in short-range forecasting, but only the two most widely used methods will be discussed here. The first of these involves the use of rules and formulae to determine the displacement and changes in intensity of such weather map features as lows, highs, fronts, waves aloft, and the jet stream. Because this method directs its efforts at specific prominent features of the weather map rather than taking into account the complete field of temperature, moisture, and momentum, the accuracy that can be achieved is limited. This technique depends greatly on highly idealized models of atmospheric phenomena, such as that of the wave cyclone discussed in Chapter 5. But characteristic lines and points such as fronts and cyclone centers do not conserve their properties with time; they are affected strongly by an environment that tends to become ever more extensive as time goes on.

The patterns of flow represented on the upper-air charts [as illustrated by Figure 7.1(b)] do have certain characteristics of behavior that are helpful to the forecaster. The sea level chart is generally dominated by several closed isobaric systems—cyclones and anticyclones—but aloft the picture changes considerably. There are few closed systems in the middle and upper troposphere; rather, a general westerly flow undulates around the poles. In middle latitudes, between 40 degrees and 50 degrees, these westerlies usually attain a peak velocity. Four or five major waves in the westerly flow pattern normally can be identified around a hemisphere, with many minor oscillations superimposed. These small undulations are associated with the fast-moving, short-lived (3 or 4 days) wave cyclones near the surface, while the longer waves are associated with the larger-scale, more sluggish features of the weather. When an area is located to the east of the ridge and west of the trough of one of these long waves, there is likely to be a 2-week or 3-week period of dry weather, while if it is to the east of the trough, wet weather is likely to prevail for a few weeks.

Prediction from equations

Meteorologists have long dreamed of being able to compute the future state of the atmosphere, much as the astronomer computes a future eclipse. The basic physical equations governing the behavior of fluids have been known for almost a century and they have been applied to the atmosphere for over 50 years. In principle, meteorologists should be able to solve these equations to provide weather forecasts. But early attempts to do this ended in dismal failure, and only recently (since about 1949) have practical, although as yet imperfect, computer forecasts been made on a regular basis.

The basic principle of large-scale numerical weather prediction is to write predictive equations for the variables that represent the weather, i.e., the horizontal wind components, the temperature, surface pressure, and water vapor content. For example, the equation for the rate of change of temperature with time at a fixed point in space is

$$\frac{\Delta T}{\Delta t} = \text{(advection + pressure change + diabatic) effects} \qquad (7.1)$$

Mathematical expressions involving the other variables are written to represent the rate of temperature change due to advection (horizontal and vertical transport of air of a different temperature), changes in pressure (expansion or compression), and diabatic effects (heating or cooling). Equations similar to (7.1) for the winds and water vapor are written for many points at different locations and levels in the atmosphere. For example, a domain the size of North America and adjacent oceans might be covered by a 50 × 50 horizontal grid at 10 levels, representing 25,000 separate points. Given an initial state of the atmospheric variables at all of these points, the time rate of change may be computed from the prognostic equations, which then yield the change of each variable at each point over a small time interval (say 1 minute). From this slightly advanced (in time) data set, the process may be repeated yielding even later estimates of the variables. By repeating this cycle hundreds of times, the horizontal and vertical distribution of the atmospheric variables can be predicted for the future. All of these calculations are performed on high-speed computers in a matter of minutes.

Although the above procedure could theoretically be carried out for months and years into the future, there are a number of serious problems that cause the forecasts to deteriorate with time. First, the mathematical equations that describe fluid motion are of a type that cannot be solved simply; so-called numerical techniques, which are extremely laborious, requiring an immense number of computations, must be used. Until the invention of high-speed computers, it was impossible to do these in a reasonable time. Second, observations exist in insufficient detail over much of the earth to permit an accurate representation of the fluid motion everywhere. Meteorologists have had to confine their solution to areas where there is a reasonable density of reports. But, of course, this introduces errors that gradually increase with time, because no portion of the atmosphere can be considered completely independent of the rest. (In other words, the weather 1000 miles away may provide the "seed" for tomorrow's weather.) Third, most of the time the *net* force acting on any part of the atmosphere is very small compared to each of the *individual* forces acting on it. For example, in the vertical, the force of gravity acting downward on 1 kilogram of air is about 9.8 newtons*; but the vertical pressure gradient force acting upward is also very nearly 9.8 newtons. This means that to compute the *net* force in the vertical, which may be only 10^{-5} newtons, one must measure the vertical pressure gradient everywhere with a very high degree of accuracy; considering techniques and density of measurement, this is impossible. Finally, there are physical difficulties, such as how to take into account the highly variable effects of friction, mountain barriers, and the distribution of heat and cold sources.

By making certain simplifications in the equations, giant computers now produce forecasts of the fields of horizontal and vertical motion on a routine daily basis. Despite the simplifying assumptions, the results are at least equal to those produced by the older, more subjective methods. Although meteorologists are still a long way from producing a weather almanac for long periods of time in advance, numerical prediction is a major step ahead in the science.

Weather forecasting

The extrapolation of prominent characteristics in the weather pattern is still the most generally used forecasting method. After the weather forecaster has decided on what the weather maps 24 to 48 hours later will look like—where the high and low centers, fronts, etc., will be located, and how intense they will be—he is still faced with the problem of relating the forecast pattern to the minutiae of "weather." Exactly where will there be precipitation and what type? Where will the clouds be and what will be the temperature over New York? To forecast these things, the meteorologist considers the factors that are likely to produce modifications in the idealized models that he has employed. For example, he must estimate whether the heating by a surface or the forced ascent of flow over a hill will be sufficient to release instability that may cause showers in an otherwise clear air mass.

The reliability of weather forecasts decreases markedly as the time interval over which they are projected increases, and so it is customary to distinguish among

*A newton (N) is the unit of force in the SI system and is equal to one kilogram · meter per second squared.

short-range (less than 48 hours), *extended-range* (up to about a week), and *long-range* (up to about a month*) forecasts. The last two are often referred to as *outlooks* and are usually quite general in character. The long-range forecasts merely state that the weather is expected to be colder or warmer, more rainy or less rainy than normal for the area and time of year.

Forecasting a Christmas Snowstorm

Throughout this book, meteorological processes have been isolated and simplified to illustrate the basic physics that affect the total weather picture. In the real world, all of the processes interact to varying degrees, and the result is an amazingly complicated evolution of atmospheric circulations on many scales. To illustrate some of the problems forecasters have when dealing with the real world, we next discuss the weather over the United States for the period 12 GMT December 24 to 12 GMT December 25, 1978. We will consider this 24-hour period from the point of view of a forecaster in three locations: Tucson, Arizona; Orlando, Florida; and Albany, New York; because the forecasting problems for these three locations were quite different on this Christmas holiday.

On the morning of December 24, 1978, a cold front extended in a north–south line from Green Bay, Wisconsin, to Baton Rouge, Louisiana, and into the Gulf of Mexico [Figure 7.1(a)]. A weak low pressure system was located on this front with center near Milwaukee, Wisconsin. To the east of this front, light southeasterly winds were bringing milder air into the Ohio Valley. To the west, temperatures in the fresh polar air mass ranged from well below zero in the north to the low 50s in southern Texas. Extending eastward along the Gulf Coast from Baton Rouge, a warm front separated truly tropical air to the south (temperatures and dewpoints in the 70s) from the old polar anticyclone to the north. As the warm moist air was lifted over the cool, dense air, heavy precipitation, and numerous thundershowers were occurring over the southeastern United States.

In the far west, the weather situation was more tranquil. A cold anticyclone was centered over Nevada, with a nearly stationary front in a trough of low pressure extending southeastward along the eastern slopes of the Rockies from Alberta into the Great Plains. As so often happens in winter, this front separated quite cold polar air in the upper Plains from milder air of Pacific origin west of the mountains. The effects of a Chinook can be seen in Nebraska. At North Platte, the temperature is −2°F and the wind is calm. Two hundred and fifty kilometers to the northwest, at Scottsbluff, gusty west winds blowing down the slopes of the Rockies are bringing warmer air to the surface, and the temperature is 33°F.

The 500-mb map for 12 GMT December 24 [Figure 7.1(b)] shows a much simpler flow pattern, with none of the local effects seen in the surface map. A large amplitude trough extends through the middle of the country, with strong

*Of course, publishers of almanacs and calendars do not hesitate to forecast a year or more in advance, and there are a few private weather forecasters who will undertake to forecast the weather many months away. However, their accuracy is generally not above that obtained from "climatology" (i.e., from predicting random variations of the average or normal weather), and if they have some scientific method, it is a well-guarded secret.

southwesterly winds to the east of the trough and northwesterly winds to the west. The jet stream enters the United States across the state of Washington, dips southeastward over Wyoming and Texas, then curves sharply back northeastward across the southeastern states. As we have seen in earlier chapters, the coldest air aloft coincides with the lowest heights in the center of the country while warmer air is located in the ridges off both coasts.

Of the three forecast locations selected for this example, it is apparent that Tucson provides the simplest situation for this day. Dominated at the surface by the large anticyclone to the north and under a warm dry ridge aloft where the air is sinking, there is no chance of precipitation for the next 24 hours. Tucson's major problem is to forecast the amount of warming today. Under clear skies and with dry surface conditions (no evaporation to hold down the maximum temperature) a rapid rise in temperature after sunrise appears likely, even though the sun will follow close to its southernmost path across the Arizona sky today. Our Tucson forecaster confidently predicts a high of 60°F.

Eastern forecasters have considerably more to worry about before they can relax and enjoy the holidays. In Orlando it is 73°F and raining. Important questions for our Orlando forecaster (and Disney World visitors) are how long the rain will last and how fast will the cold front to the west advance. Noting the dry weather at Tampa and other Florida stations to the west [Figure 7.1(a)], the forecaster says the current rain will be ending soon. However, with the upper-level trough and cold front advancing from the west, more showers and thunderstorms are expected later in the day. Upper-level (500-mb) winds are from the west at 50 knots (93 km/h) and the surface front is currently about 1000 km to the west of Orlando. If the front moves at the speed of the 500-mb wind, it will take about 11 hours to reach Orlando. However, surface weather systems usually move at a somewhat slower speed than the upper-level wind; this front has been moving at about 30 knots (55 km/h). Thus our Orlando forecaster predicts frontal passage in about 18 hours, or 06 GMT (around midnight local time). Significantly cooler and drier air is therefore in the offing for Christmas day.

The most challenging forecast is for the northeast. The advancing complex surface low pressure system and the 500-mb trough promise precipitation later in the day. The main questions are where the precipitation will be rain and where it will be snow, and then how much precipitation will fall. At 12 GMT December 24, temperatures are below freezing north of a line extending from Philadelphia to Milwaukee. Warm southerly winds ahead of the advancing low are pushing this line slowly northward. If the Wisconsin low remained the major storm and moved east–northeast, it would pass through southern Ontario, bringing warm air to all points south of its path. In this case, rain would fall over Pennsylvania, New York, and most of New England, and the storm would be relatively harmless.

Disturbing signs are present, however, in the surface and upper-air maps (Figure 7.1) and indicate the possibility of a quite different scenario. Heavy percipitation and large pressure falls in the southeast indicate strong upper-level divergence and upward motion here. At 500 mb [Figure 7.1(b)], the strongest winds are over the Gulf Coast, indicating a more favorable environment for the southern storm to develop compared to the northern storm. If the low over Alabama did intensify, it would follow the southwesterly winds aloft and move up the East Coast. The warm advection over the northeast would end as northeasterly winds developed in advance of the new storm. Heavy snow would be almost certain north of Pennsylvania, with snow possible in the mountains of Pennsylvania and West Virginia.

A major factor in the Albany forecaster's final decision on this day was the computer model's forecast. After performing millions of calculations, the results were clear. The southern low would develop rapidly following the demise of the northern cyclone, and a major winter storm was likely for the northeast. Heavy snow warnings were issued for all of New England. However, even with the major question resolved, forecasts for cities along the boundary between subfreezing and above freezing temperatures remained a problem. The heaviest snow often falls along the line separating rain and snow, and the mislocation of this line by 100 km could make the difference between Harrisburg, Pennsylvania, receiving an inch of rain or a foot of snow. Not even the computer models could produce such a detailed forecast. Making the best estimate possible, the forecasters put the rain–snow boundary through Erie, Pennsylvania, eastward to New York. The exact location would depend on how strong the southern storm became.

In the next 12 hours the weather behaved pretty much as expected. By 00 GMT December 25 [Figure 7.2(a)] the temperature in Tucson rose to 61°F as forecast. In the southeast, the cold front was making steady progress eastward, ending the rain along the Gulf Coast. Ahead of the front, scattered showers and thunderstorms were occurring over Florida; but the biggest weather story was the rapidly moving east coast storm, which had intensified from a minimum pressure of 1002 mb 12 hours earlier to its present central pressure of 992 mb. Heavy precipitation spread rapidly northeastward during the day, with the rain–snow line holding through central Pennsylvania. At 500 mb [Figure 7.2(b)] the trough shifted rapidly eastward. The jet stream over the southeast increased from a maximum of 85 knots [Figure 7.1(b)] to 110 knots [Figure 7.2(b)] as the cyclone intensified.

Figure 7.4a shows a photograph in the visible wavelengths made by the geostationary satellite for 1930 GMT December 24, 4½ hours earlier than the surface map of Figure 7.2(a). Clouds cover most of the eastern United States. The deepest and most active clouds, as indicated by the irregular appearance of the tops, occur over the Carolinas and Virginias. This rough appearance is caused by the highest cloud elements casting shadows on the lower clouds. A line of showers and thunderstorms originates in the Gulf of Mexico and crosses central Florida, marking the leading edge of the advancing cold front.

Figure 7.4(b) shows a photograph of the developing storm taken in the infrared wavelengths. The radiation seen by the satellite is converted to a temperature, which is converted to a shade of gray ranging from black to white. Horizontal variations of this gray scale indicate horizontal variations in temperature. Thus the warm Atlantic Ocean surface in the southeast appears uniformly dark; strong contrasts in gray scale occur along the edges of the cloud patterns such as along the frontal band of clouds over Florida. The massive cloud cover over New England appears as a light gray area, indicating cold cloud tops.

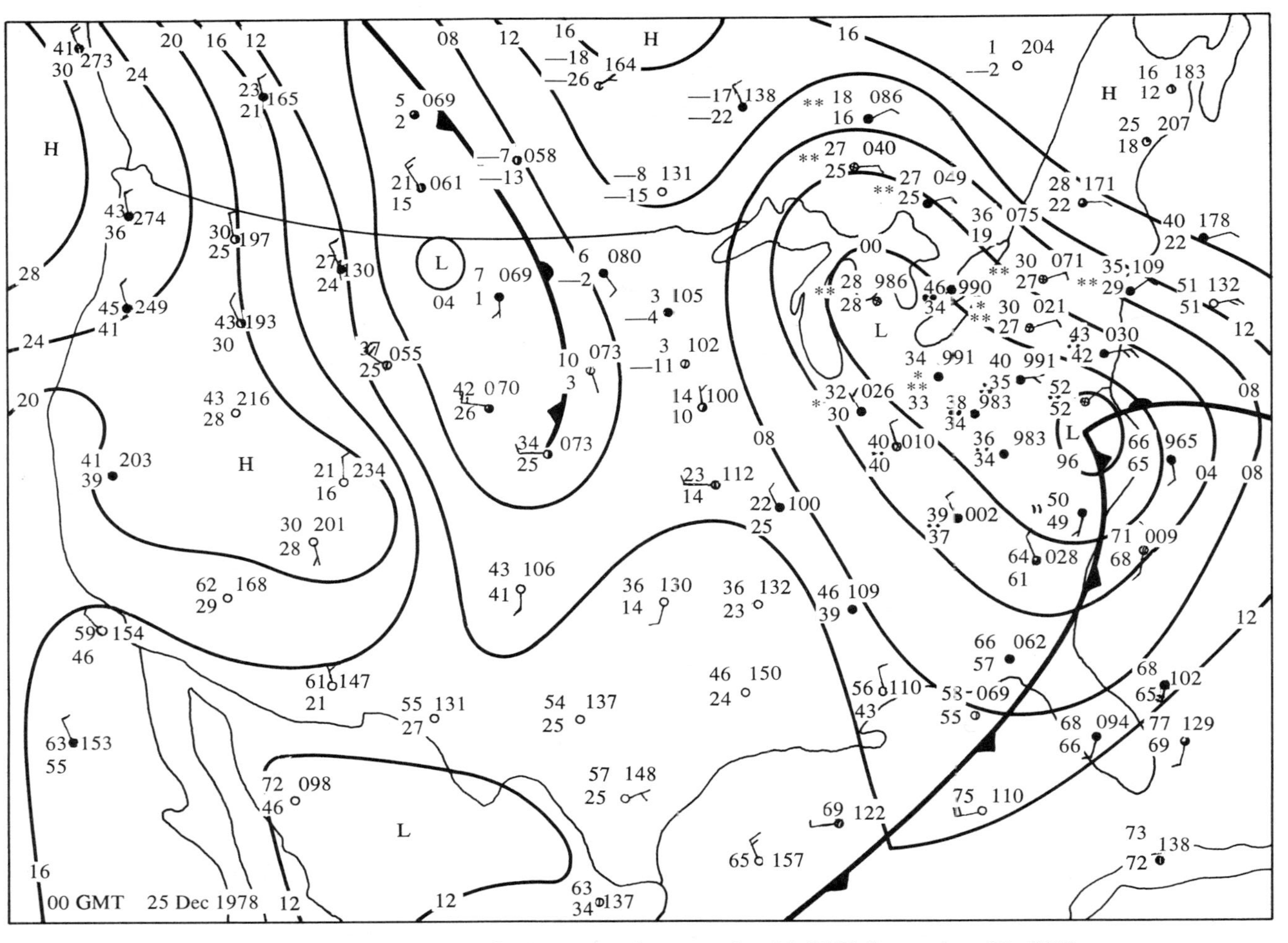

FIGURE 7.2(a) *Surface weather map for 00 GMT December 25, 1978.*

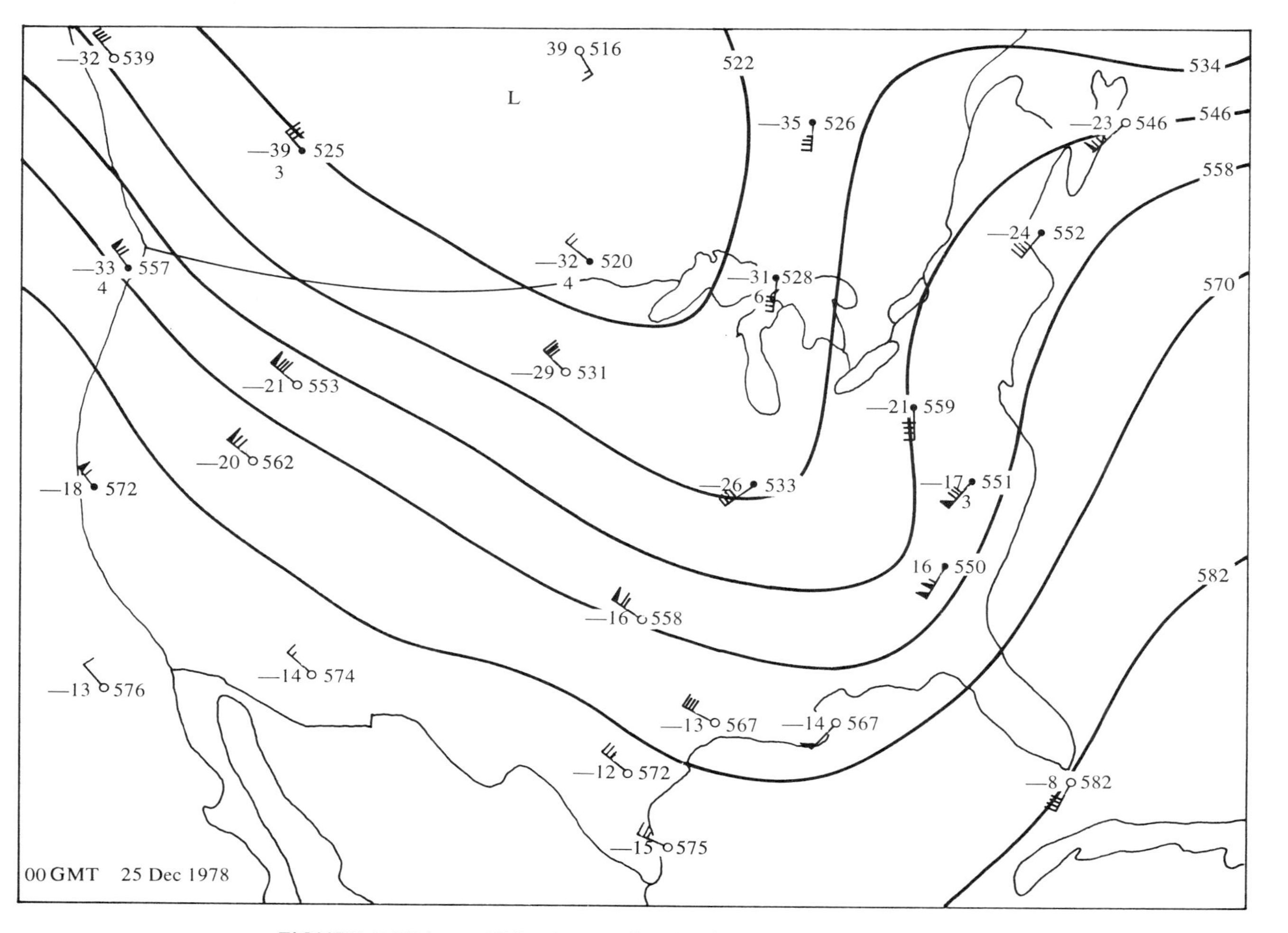

FIGURE 7.2(b) *500-mb map for 00 GMT December 25, 1978.*

With the evolution of the storm well underway, the forecast for the next 12 hours for the northeast was straightforward. Heavy snow was a certainty for New England, with the entire northeast becoming windy and colder after the storm passed and winds shifted into the northwest. For Orlando, the cold front would pass in the next 12 hours ending the threat of showers there. At Tucson, strong radiational loss of surface heat through a dry atmosphere would lower temperatures by 30°F or more by morning, and scattered frost was a possibility.

By 12 GMT December 25, the east coast cyclone had deepened another 10 mb, reaching a minimum pressure of 982 mb. Children from the mountains of West Virgina northward into western Pennsylvania and all of New England except the coastal areas woke to find a white Christmas. Over 40 cm of snow had fallen in upper New York from the storm, which was now at its peak intensity over Massachusetts. South of the storm, westerly winds swept the cold front into the Atlantic, ending the showers and thunderstorms along the east coast. At Orlando, the temperature and dewpoint were nearly 20°F lower than they were at the same time yesterday [compare Figures 7.3(a) and 7.1(a)]. In Tucson, the temperature at the instrument shelter was 39°F; however, frost covered cars and low grassy surfaces which had cooled by radiation to temperatures below the frost point.

At 500 mb [Figure 7.3(b)] the trough axis was continuing to rotate eastward. Over Pittsburg the height of the 500-mb surface had dropped from 5520 to 5280 m, while the temperature fell 5°C in the last 24 hours [compare Figures 7.3(b) with 7.1(b)]. The jet stream continued to intensify as potential energy was converted to kinetic energy; the winds over Atlanta were 130 knots this morning.

Figures 7.5(a) and (b) show visible and infrared satellite photographs for 1430 GMT December 25, 1978. The massive swirl of deep clouds has moved into Canada with clearing occurring over the southeast and downwind of the Appalachian Mountains. Over the mountains of West Virginia, Virginia, and Pennsylvania, however, low clouds persist with parallel bands of clouds occurring over the individual ridges. The band of clouds associated with the cold front has weakened considerably [compare with Figures 7.4(a), and (b)] and has moved well off the United States' coast.

Thus for this 24-hour period all three forecasts worked out satisfactorily. The origin of the east coast cyclone was adequately resolved by the upper-air network over the southern United States and the computer model faithfully simulated the major features of the storm. The forecasters adequately filled in the details and tailored the forecasts to local effects such as variations in elevation.

Extended and long-range forecasts

Extended-range forecasting is most commonly accomplished through analysis of the slowly changing, large-scale features of the atmospheric circulation. By averaging the daily charts at various levels over several days, the smaller, fast-moving "eddies" and waves in the flow are suppressed and only the persistent large-scale features remain. Apparently, these big undulations of the averaged flow determine to a great extent the character of the weather over a week or two, just as the migratory wave cyclones produce many of the day-to-day changes at middle and high latitudes.

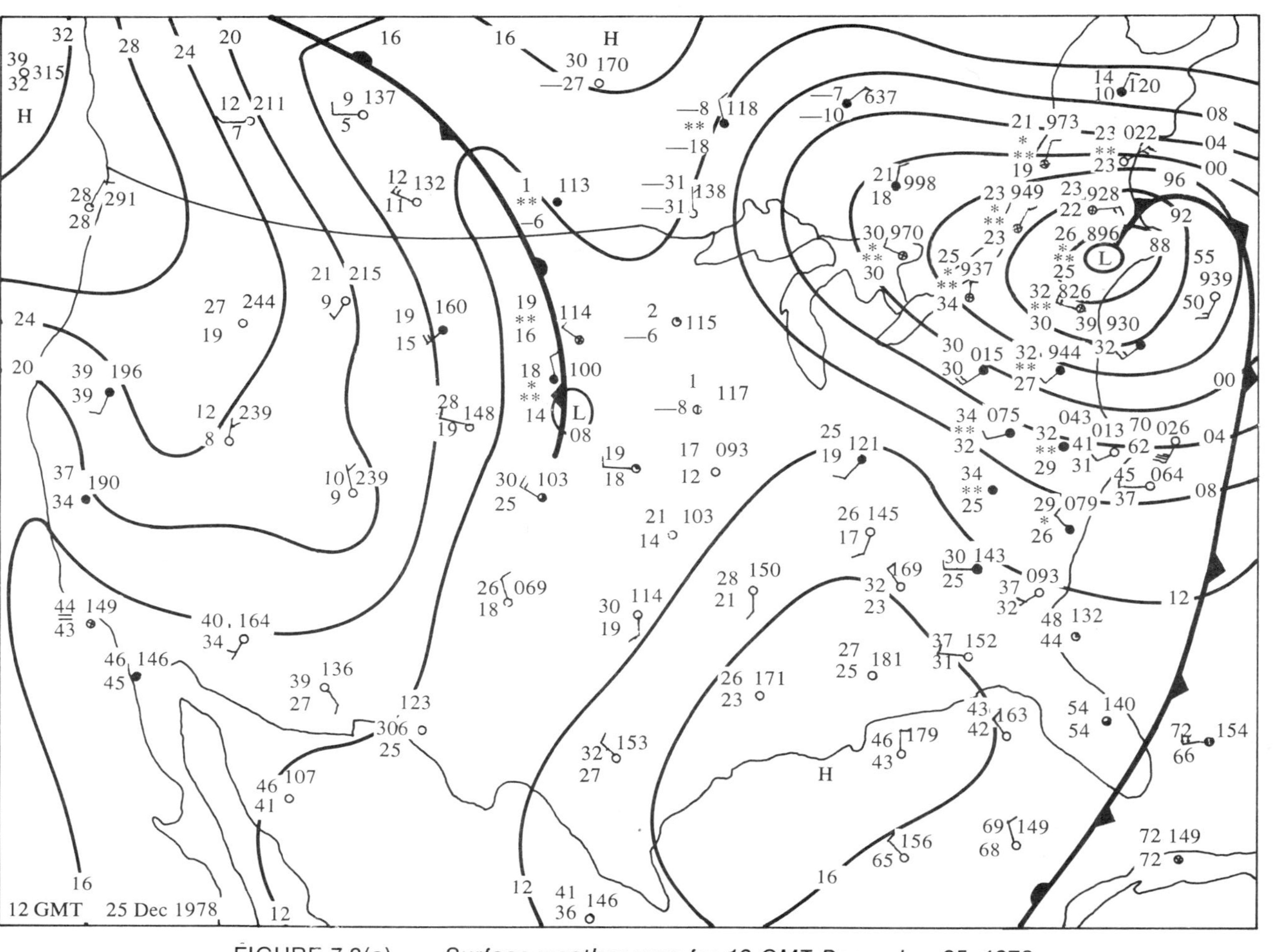

FIGURE 7.3(a) *Surface weather map for 12 GMT December 25, 1978.*

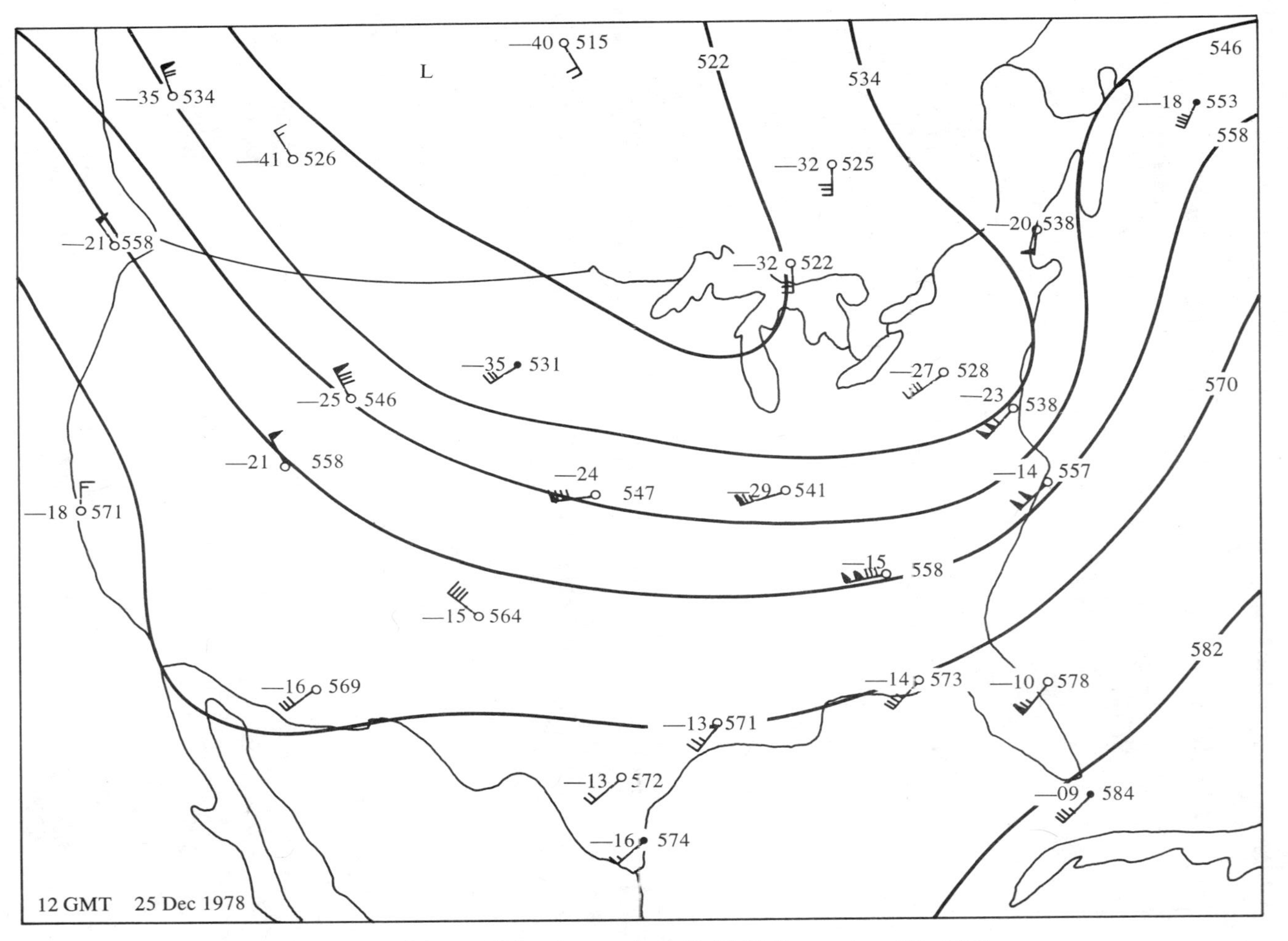

FIGURE 7.3(b) *500-mb map for 12 GMT December 25, 1978.*

FIGURE 7.4(a) *Visible satellite photograph for 1930 GMT December 24, 1978.*

For forecasts beyond 30 days, all present methods are based on statistical analysis. The search for periodicities (cycles) in weather elements seems to be a favorite occupation of many amateurs and professionals. But aside from the well-known daily and annual cycles caused by the earth's movements, none seems to be very reliable for forecasting. Another technique that is used is that of *analogues;* it consists of looking for weather patterns from past records that resemble the present one and then forecasting the present situation to evolve in the future in the same way as did the analogous one in the past. Until more is learned about what controls the average large-scale flow in the atmosphere—the general circulation—long-range forecasting cannot have a solid physical foundation.

Where an operation requires that plans be made for months or years in advance, useful weather information can be obtained from climatological data. Such information as the average and extremes of weather conditions for past years, as well as the

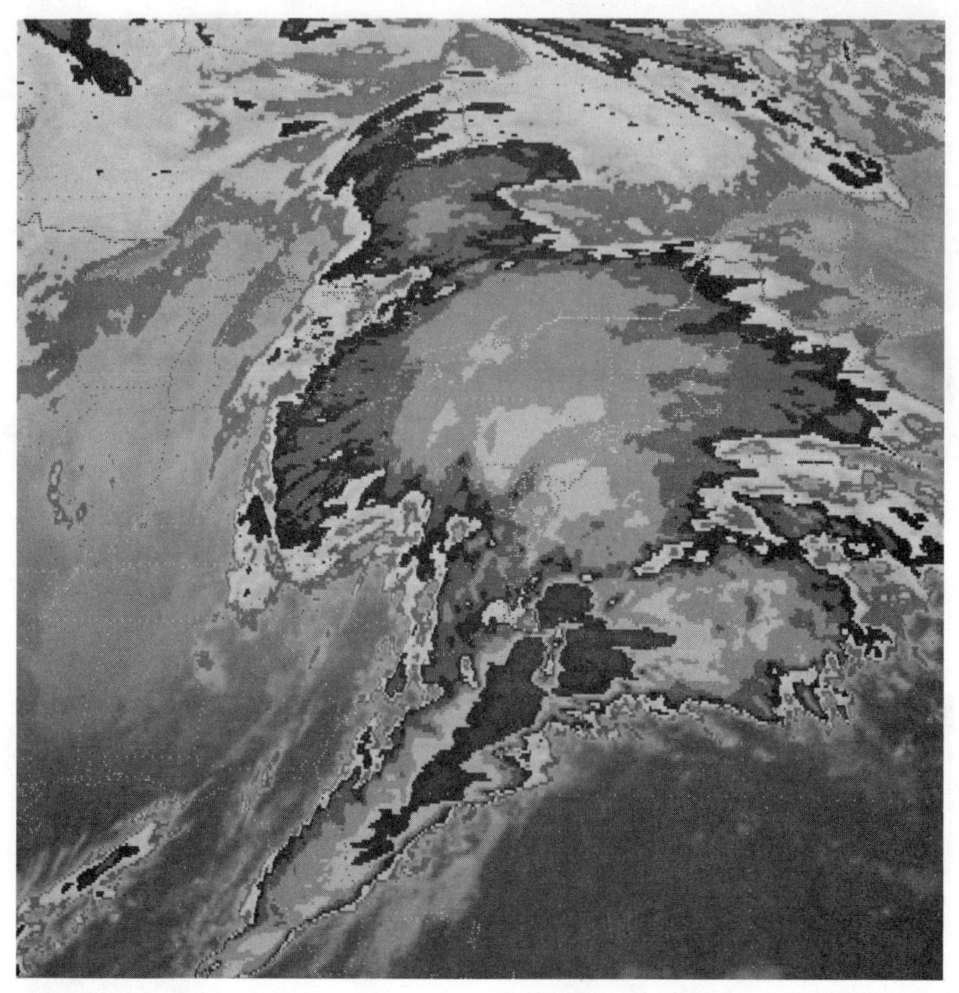

FIGURE 7.4(b) *Infrared satellite photograph for 2230 GMT December 24, 1978.*

frequency of occurrence of operationally critical conditions, can provide helpful advice for all kinds of long-range planning.

7.3 Weather Modification

Humans have been modifying their atmospheric environment since they first lit fires and moved into caves. But even outside of his shelter, man has been modifying his weather, both intentionally and unintentionally, for a long time. He has done it by changing the contours of the land, changing the surface properties, and contaminating the air. But all of these changes have been done on a relatively small scale.

The various scales of the weather patterns must be kept in mind when one considers the feasibility of any particular scheme for deliberate weather modification. On

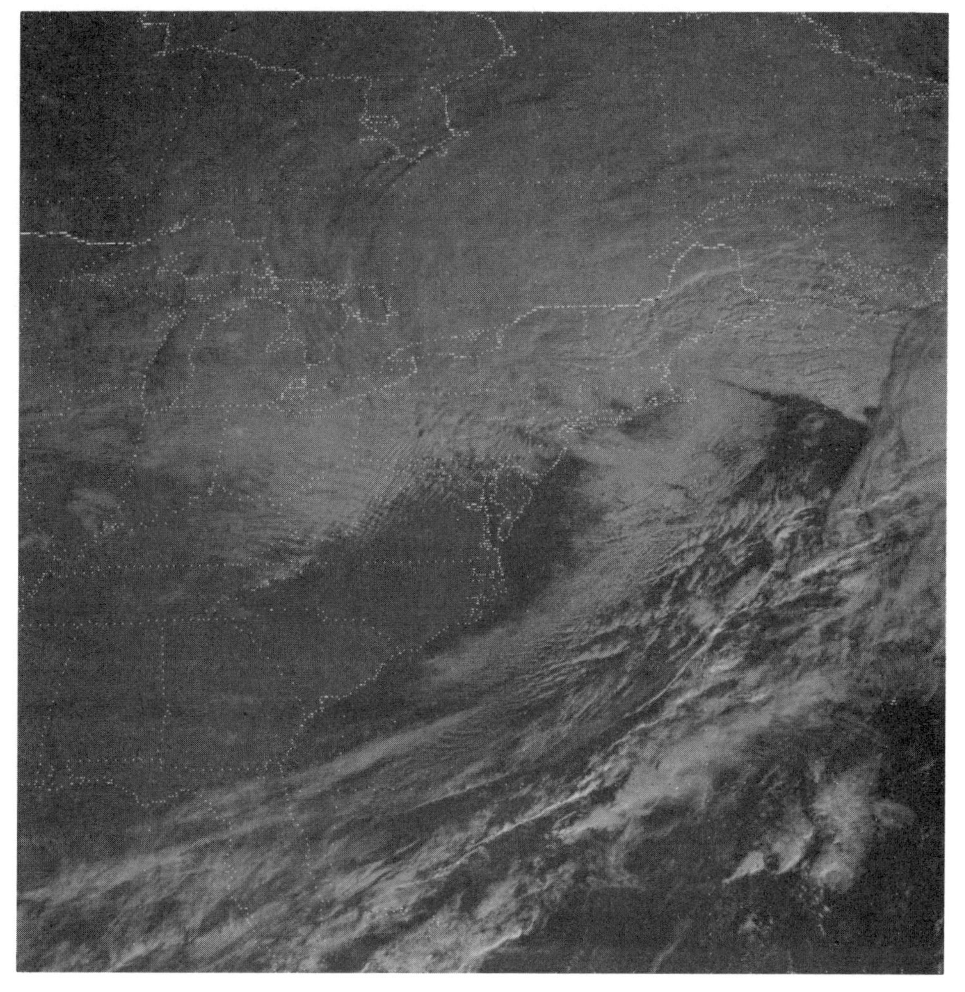

FIGURE 7.5(a) *Visible satellite photograph for 1430 GMT December 25, 1978.*

a large or medium scale, we cannot hope, at least in the foreseeable future, to change the climate by inputs of energy equaling those of natural atmospheric processes. We could hardly match the rate at which heat energy is converted to kinetic energy even in a small thunderstorm, and the total rate of energy conversion increases greatly as the circulation size increases. (See Problem 1, Chapter 5.)

On a small scale, there are a few examples of the direct use of heat energy to change the weather. Heating of the air over crops to save them from frost damage has been practiced for a long time. Generally, however, the produce must have a fairly high value, such as citrus fruit has, to make the practice economically feasible. During World War II, fog over English airports was sometimes dissipated sufficiently to allow aircraft operations by burning oil to raise the air temperature locally a few degrees above the dew point. In some places sidewalks are kept clear of snow by running hot water or steam pipes beneath the pavement. In the densely populated

FIGURE 7.5(b) *Infrared satellite photograph for 1400 GMT December 25, 1978.*

areas of New York City during the winter, the thermal energy released to the atmosphere while heating buildings and generating power actually exceeds the amount received from the sun at ground level. But in these examples, we are dealing with releases of rather large amounts of heat energy over relatively small areas. These energies do affect the climatic conditions of these limited areas. But when considered over larger scales, man's input of energy into the atmosphere is, at present, minuscule. For example, the total electrical energy produced each day in the United States is equivalent to just the latent heat energy that is released during precipitation of 2.5 millimeters (0.1 inch) over an area of 50 kilometers square (an area about half the size of Rhode Island).

Even though it seems unlikely that man has used or soon will use "brute force" to modify his atmospheric environment, except on a very small scale, we might change the weather in either or both of two ways: (1) by altering the existing "natural" energy exchanges that occur among the earth's surface, the atmosphere, and space, or (2) by stimulating or "triggering" various forms of instability that sometimes arise in atmospheric processes. These two "methods" are not mutually exclusive, of course, since a change in the energy balance may trigger instabilities and the setting off of instabilities may release significant amounts of energy. An example of interference with the natural energy balance is the change that occurs when a city is built on what were green fields. (Recall the climatic differences between the city and the countryside, discussed in Chapter 6.) An example of triggering atmospheric processes is cloud seeding: super-cooled water droplets in a cloud may be induced to freeze, grow at the expense of the remaining water drops in the cloud, and precipitate; at the same time, the release of the latent heat of fusion may provide energy for additional vertical growth of the cloud.

Most weather changes that have been produced or proposed are those that result from changes induced in the composition of the air or clouds, or in the surface properties of the earth. One example of a change in surface properties that produces a change in the microclimate is that caused by the extensive areas of asphalt and concrete in a city. Water is often used to control the temperature over crops. On a clear, cold night, the temperature over an irrigated field will be noticeably higher than over dry soil.

Pollutants in the air affect the heat content of the atmosphere by reducing its transparency to the sun's income energy and the earth's outgoing heat. Since dust particles serve as cloud nuclei, pollution leads to denser fogs and smogs. Fog is much more frequent over cities than over the surrounding countryside. It may also be that the greater number of nuclei leads to more precipitation over cities.

Modification of clouds by use of dry ice and silver iodide has already been mentioned in Chapter 1. Cloud seeding has not been proved to produce significant changes in precipitation amounts, except under certain favorable circumstances over relatively small areas. In fact, it has been much more effective in the *dissipation* of cold (less than 25°F) fogs than as a means for stimulating the rainfall rate. Figure 7.6 shows the effects of seeding a cloud from the air.

Control over the water supply can be achieved in ways other than by increasing precipitation. Suppression of evaporation from lakes and reservoirs is one technique that has been employed. This has been done by spreading a monomolecular film of a substance such as acetyl alcohol on the surface of the water. Evaporation can be retarded by 15–20 percent in this way, but it is difficult to prevent the film from being broken by waves.

Snow is a very important natural "reservoir" of water for many places. If the rate at which snow melts could be controlled, a steady supply of water might be available throughout the year instead of having an overbalance in the spring. Increasing the rate of snowmelt is not too difficult. The high reflectivity of snow can be decreased by covering the surface with some dark material such as lampblack. In Tibet it has long been the practice to throw pebbles on snow fields to speed melting for early planting. But no practical method for retarding snowmelt on a large scale has yet been suggested; slowing snowmelt would be immensely valuable to regions such as California, where most of the year's water supply comes in the form of snow over the mountains.

FIGURE 7.6 *Cloud deck seeded from the air. Note where cloud has been dissipated by seeding. (Courtesy of AFCRL.)*

Various proposals have been made for weather modification on a grandiose scale. For example, it has been suggested that large areas of lowlands be flooded to temper the climate of surrounding areas. In northern Siberia the enormous annual temperature range (over 100°F) might be sharply reduced by extensive flooding.

A land bridge across the Bering Strait, cutting off the circulation of waters of the Arctic and Pacific Oceans, has existed in the past, and it has been proposed that this could be rebuilt. This barrier would stop the flow of heat between these bodies of water, presumably raising the temperatures to the south of the strait and increasing the temperature contrast across the barrier.

The question of weather control is a very important one and deserves serious consideration. But until the atmospheric scientist understands atmospheric processes more fully and therefore the possible effects that his tinkering may have, he must proceed cautiously. After all, man is very delicately tuned to his existing environment.

PROBLEMS

1. Identify the major ridges and troughs at 500 millibars in Figures 7.1–7.3. How do the sea-level cyclones and anticyclones move in relation to the large-scale wave patterns?
2. Determine the speed and direction of displacement of the cyclone on the sea-level charts of Figures 7.1–7.3 for the two 12-h periods (12 GMT Dec. 24 to 00 GMT Dec. 25 and 00 GMT Dec. 25 to 12 GMT Dec. 25).
3. Is the complexity of the frontal analyses on the sea level charts greater over the oceans or over the continents? Why should there be a difference?

4. Examine the latest weather map published in your local newspaper. Predict whether there will be precipitation (and, if so, the type) and what the temperature, wind, and cloudiness will be at your city 24 hours from the time of the map. List the factors you took into account in predicting each element.
5. Some proverbs state that physical appearance of certain insects and animals is an indication of future weather. Do you doubt their validity? Why?
6. Compare the total energy received from the sun within the boundaries of your city on January 1 with that generated by power plants and heating units. (Use Fig. 3.7.)
7. List all the ways in which you think man may be affecting the climate on various scales, both intentionally and unintentionally. Can you find any evidence from weather records that the microclimate in your area has been changed by man?
8. Why is a "ring around the moon" a fairly good indication of an approaching storm? Can you think of a reason why the following weather rhyme might have a sound meteorological basis?

   Rainbow in the morning, sailor's warning
   Rainbow at night, sailor's delight.

9. During the middle of the afternoon on a quiet summer day, the temperature of the air is 82°F, the dew point is 57°F, and cumulus clouds are observed. How high are the bases of the clouds? What is the temperature at the cloud bases? What would be the height of the freezing level in the clouds (assuming that they extend that high)?
10. List all the ways in which you think man may be affecting the climate on various scales (i.e., from areas as small as, say, your backyard, to regions as large as an entire continent, both intentionally and unintentionally). Can you find any evidence from weather records that the microclimate in your area has been changed by man?
11. How did the temperature, winds, and heights change at 500 mb over Peoria, Illinois, between 12 GMT Dec. 24 and 12 GMT Dec. 25, 1978? [Compare Figures 7.1(b) and 7.3(b).]

Appendix 1

Units Used in this Book

1. Temperature Scales

	Fahrenheit (F)	*Celsius (C)*	*Kelvin (K) or Absolute (A)*
Boiling point of water	212	100	373
Melting point of ice	32	0	273
Divisions between fixed points	180	100	100

Conversion formulas: $\frac{°F - 32}{180} = \frac{°C}{100}$; $K = °C + 273°$

2. Length

1 kilometer = 0.6214 statute mile = 0.5396 nautical mile
1 meter(m) = 1.093611 yards = 3.2808 feet = 39.370 inches
1 cm = 0.3937 in. = 10^4 micrometers (μm) = 10^8 angstroms (Å)
1 mile = 1.61 km

3. Velocity

1 knot (nautical mile per hour) = 1.1516 statute mi/h = 0.5148 m/s
1 mi/h = 0.8684 knot = 0.447 m/s
1 m/s = 2.2369 mi/h = 1.9424 knots = 3.2808 ft/s
1 km/h = 0.62 mi/h

4. Force

Newton(N) = 1 kg-m/s^2
1 dyne = 10^{-5} N
1 pound = 4.4482 N
(British)

5. Pressure

1 Pascal (Pa) = 1 newton/m^2 (N/m^2) = 10^{-2} millibar (mb)
1 mb = 0.02953 inches of mercury (in Hg) = 1000 dynes/cm^2
1 in. Hg = 33.8639 mb
1 lb/in^2 = 68.947 mb

6. Energy

1 gram-calorie [or, just "calorie," (cal)]
1 erg = 1 dyne cm = 2.388×10^{-8} cal
1 watt-hour = 860 gram-calories (g-cal) = 3.600×10^{10} ergs
1 British thermal unit (Btu) = 0.293 watt-hour = 251.98 gram-calories = 1.055×10^{10} ergs
1 joule (J) = 10^7 ergs
1 cal = 4.1855×10^7 ergs
1 foot-pound = 1.356×10^7 ergs
1 horsepower-hour = 2.684×10^{13} ergs = 0.6416×10^6 cal

7. Power

1 watt = 14.3353 cal min^{-1}
1 cal min^{-1} = 0.06976 watt
1 horsepower = 746 watts
1 Btu min^{-1} = 175.84 watts = 252.08 cal min^{-1}

8. Powers of Ten

⋮
$10^{-2} = 0.01$
$10^{-1} = 0.1$
$10^0 = 1.0$
$10^1 = 10$
$10^2 = 100$
⋮

9. Prefixes

Factors by which unit is multiplied	Prefix	Symbol
10^{12}	tera	T
10^9	giga	G
10^6	mega	M
10^3	kilo	k
10^2	hecto	h
10	deka	da
10^{-1}	deci	d
10^{-2}	centi	c
10^{-3}	milli	m
10^{-6}	micro	μ
10^{-9}	nano	n
10^{-12}	pico	p
10^{-15}	femto	f

Appendix 2

Standard Atmosphere

Altitude (*m*)	*Temperature* (°*C*)	*Pressure* (*mb*)	*Density* (kg/m^3)
0	15.0	1,013.2	1.2250
500	11.8	954.6	1.1673
1,000	8.5	898.8	1.1117
1,500	5.2	845.6	1.0581
2,000	2.0	795.0	1.0066
2,500	− 1.2	746.9	0.9569
3,000	− 4.5	701.2	0.9092
3,500	− 7.7	657.8	0.8634
4,000	−11.0	616.6	0.8194
4,500	−14.2	577.5	0.7770
5,000	−17.5	540.5	0.7364
5,500	−20.7	505.4	0.6975
6,000	−24.0	472.2	0.6601
6,500	−27.2	440.8	0.6243
7,000	−30.4	411.0	0.5900
7,500	−33.7	383.0	0.5572
8,000	−36.9	356.5	0.5258
8,500	−40.2	331.5	0.4958
9,000	−43.4	308.0	0.4671
9,500	−46.7	285.8	0.4397
10,000	−49.9	265.0	0.4140
10,500	−53.1	245.4	0.3886
11,000	−56.4	227.0	0.3648
11,100	−56.5	223.5	0.3593
11,500	−56.5	209.8	0.3374
12,000	−56.5	194.0	0.3119
13,000	−56.5	165.8	0.2666
14,000	−56.5	141.7	0.2279
15,000	−56.5	121.1	0.1948
16,000	−56.5	103.5	0.1665
17,000	−56.5	88.5	0.1423
18,000	−56.5	75.6	0.1216
19,000	−56.5	64.7	0.1040
20,000	−56.5	55.3	0.0889
25,000	−51.6	25.5	0.0401
30,000	−46.6	12.0	0.0184
35,000	−30.6	5.7	0.0085
40,000	−22.8	2.9	0.0040
45,000	− 9.0	1.5	0.0020
50,000	− 2.5	0.8	0.0010
60,000	−17.4	0.225	0.000306
70,000	−53.4	0.055	0.000088
80,000	−92.5	0.010	0.000020
90,000	−92.5	0.002	0.000003

Appendix 3

Plotting Model For Sea Level Weather Chart

WW Symbols

- haze
- fog
- drizzle
- rain
- snow
- rain shower
- thunderstorm
- snow shower

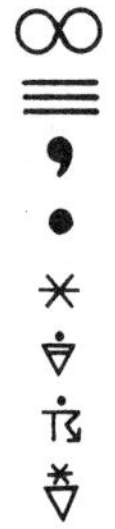

Symbols	*Example*
N: Amount of total sky cover	Overcast
ff: Barbs show wind speed (full barb = 10 knots)	25 knots
dd: Arrow shaft shows wind direction	Northwest
TT: Temperature (°F)	38°F
VV: Visibility (miles)	1.5 miles
WW: Weather type	Continuous light rain
T_dT_d: Dew point temperature (°F)	34°F
C_l: Type of low clouds	stratus
h: Height of ceiling	300–599 ft
N_h: Amount of low cloud cover	6/8
RR: Precipitation amount, past 6 hours	0.26 in.
W: Weather, past 6 hours	Rain
R_t: Time precipitation began or ended	Began 3–4 hours ago
a: Trend of barograph curve, past 3 hours	Rising
pp: Pressure change, past 3 hours	+2.8 mb
PPP: Sea level pressure, with only last three digits (including tenths) given	1,013.2 mb
C_m: Type of middle clouds	Nimbostratus
C_h: Type of high clouds	Cirrus

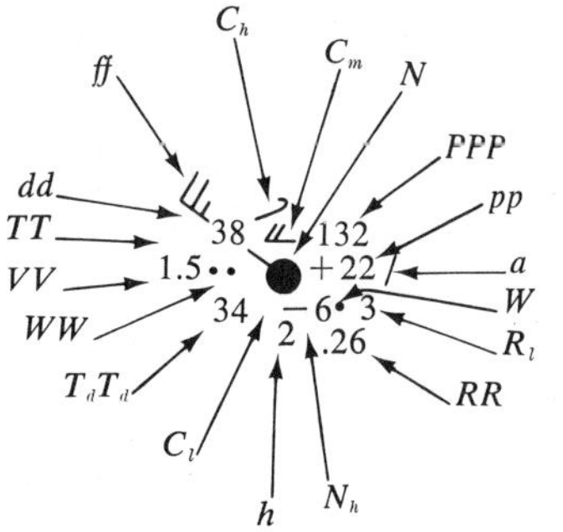

Plotting Model For Upper-air Chart

Symbols	*Example*
dd: Arrow shaft shows wind direction	270°
ff: Barbs show wind speed (triangle, 50 knots; full barb, 10 knots)	65 knots
TT: Temperature (°C)	5.3°C
T_dT_d: Temperature − dewpoint difference (°C)	−1.6°C
hhh: Height of pressure surface (in meters), with only first three digits given	5400 m (500 mb surface)

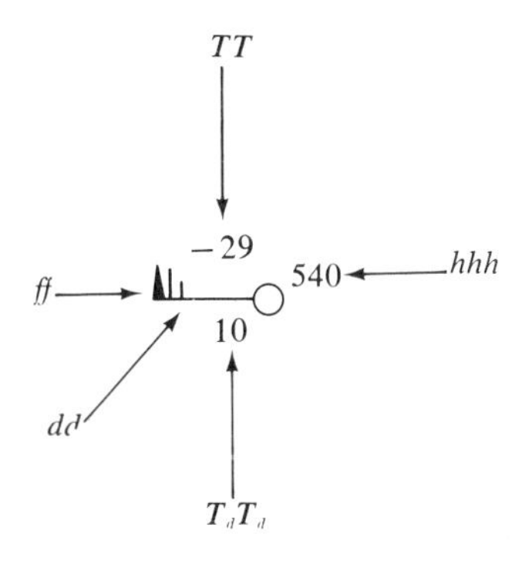

Appendix 5

Supplementary Readings

Elementary books on general meteorology and climatology

Anthes, Richard A., Hans A. Panofsky, John J. Cahir, and Albert Rango, *The Atmosphere*. Columbus, Ohio: Charles E. Merrill Publishing Co., 1975.

Battan, Louis J. *The Nature of Violent Storms*. Garden City, N.Y.: Doubleday & Co., Inc., 1961.

Blair, Thomas A. and Robert C. Fite, *Weather Elements*. Englewood Cliffs, N.J.: Prentice-Hall, Inc., 1965.

Brooks, C. E. P., *Climate in Everyday Life*. New York: Philosophical Library, Inc., 1951.

Dobson, G. M. B., *Exploring the Atmosphere*. New York: Oxford University Press, 1963.

Hare, F. K., *The Restless Atmosphere*. New York: Harper & Row Publishers, Inc., 1963.

Mason, Basil J., *Clouds, Rain and Rainmaking*. New York: Cambridge University Press, 1962.

Miller, Albert and Jack C. Thompson, *Elements of Meteorology*. Columbus, Ohio: Charles E. Merrill Publishing Co., 1970.

Petterssen, S., *Introduction to Meteorology*. New York: McGraw-Hill Book Co., Inc., 1969.

Scientific American, Offprint Series. San Francisco, California: W. H. Freeman and Co. The following offprints are especially useful:

612. McDermott, Walsh, *Air Pollution and Public Health*.
823. Plass, Gilbert N., *Carbon Dioxide and Climate*.
824. Landsberg, Helmut E., *The Origin of the Atmosphere*.
841. Starr, Victor P., *The General Circulation of the Atmosphere*.
847. Malkus, Joanne Starr, *The Origin of Hurricanes*.
848. Tepper, Morris, *Tornadoes*.
876. Myers, Joel N., *Fog*.
881. Stewart, R.W., *The Atmosphere and the Ocean*.

Spar, Jerome, *Earth, Sea, and Air. A Survey of the Geophysical Sciences*. Reading, Mass.: Addison-Wesley Publishing Co., 1965.

Sutton, O. G., *The Challenge of the Atmosphere*. New York: Harper & Bro., 1961.

Trewartha, Glenn T., *An Introduction to Climate*. New York: McGraw-Hill Book Company, Inc., 1968.

Some sources of weather data

The Smithsonian Institution, Washington, D.C. "World Weather Records"
U.S. Weather Service (NOAA), Washington, D.C.:

Average Monthly Weather Resume and Outlook
Climates of the States
Climatic Charts for the United States
Climatological Data for the U.S. by Sections
Daily Weather Map
Monthly Climatic Data for the World

(In addition, the U.S. Weather Service publishes many pamphlets on particular meteorological phenomena.)

Index